RANDOM PROCESSES IN LINEAR SYSTEMS

MICHAEL B. PURSLEY

Holcombe Professor of Electrical
and Computer Engineering
Clemson University

Prentice Hall, Upper Saddle River, New Jersey 07458

Library of Congress Cataloging-in-Publication Data
Pursley, Michael B.
    Random processes in linear systems / by Michael B. Pursley.
        p. cm.
    Includes bibliographical references and index.
    ISBN 0-13-067391-9
    1. Stochastic processes. 2. Linear systems. I. Title.

QA274 .P87 2002
519.2--dc21

                                          2001050071

Vice President and Editorial Director, ECS: *Marcia J. Horton*
Acquisitions Editor: *Eric Frank*
Editorial Assistant: *Jessica Romeo*
Vice President and Director of Production and Manufacturing, ESM: *David W. Riccardi*
Executive Managing Editor: *Vince O'Brien*
Managing Editor: *David A. George*
Production Services/Composition: *Preparé Inc.*
Director of Creative Services: *Paul Belfanti*
Creative Director: *Carole Anson*
Cover Designer: *Jayne Conte*
Art Editor: *Greg Dulles*
Manufacturing Manager: *Trudy Pisciotti*
Manufacturing Buyer: *Lynda Castillo*
Marketing Manager: *Holly Stark*

 © 2002 by Prentice Hall
Prentice-Hall, Inc.
Upper Saddle River, New Jersey 07458

Printed in the United States of America
10 9 8 7 6 5 4 3 2 1

ISBN 0-13-067391-9

Pearson Education Ltd., *London*
Pearson Education Australia Pty. Ltd., *Sydney*
Pearson Education Singapore, Pte. Ltd.
Pearson Education North Asia Ltd., *Hong Kong*
Pearson Education Canada, Inc., *Toronto*
Pearson Educacíon de Mexico, S.A. de C.V.
Pearson Education—Japan, *Tokyo*
Pearson Education Malaysia, Pte. Ltd.
Pearson Education, *Upper Saddle River*, *New Jersey*

To my mother, Evelyn Pursley,
the memory of my father, Bader Pursley,
my wife, Lou Ann, and our daughter, Jessica

Contents

Preface

This book provides an introduction to random processes that is suitable for advanced undergraduates, beginning graduate students, and practicing engineers. Portions of the book have been used as a text for 12 to 15 hours of instruction in the second part of an undergraduate course in probability and random processes at the University of Illinois. The material was taught by the author and other faculty members to juniors and seniors majoring in electrical engineering or computer engineering. The course emphasized Chapters 2 and 3, but it included parts of Chapter 4. For thorough coverage of Chapter 4, more time is required. The manuscript for the book has also served as the text for the first few weeks of several graduate courses in communications at a number of universities. In most of these courses, Chapters 2–4 are covered completely. In addition, the manuscript has been used by many beginning graduate students who needed to acquire a solid understanding of second-order random processes in preparation for graduate courses in digital communications, wireless communication systems, or signal processing. Typically, such a student's undergraduate education included basic probability and linear systems, but not random processes.

The book presents a very brief review of probability, but this material is intended primarily to introduce the notation for subsequent chapters. The student is expected to have a good understanding of probability and random variables from a previous course on probability or from the earlier portion of a course on probability and random processes. An adequate understanding of the concepts needed for the student to begin a study of random processes typically requires approximately 25 to 30 hours of instruction based on a text such as *A First Course in Probability* by Sheldon Ross or *Introduction to Probability and Its Applications* by Richard Scheaffer. Such instruction might be given in the first part of a course for which this book serves as a text for the second part.

There is enough material here for a separate one-quarter course in random processes at the undergraduate or beginning graduate level. For a graduate course it is anticipated that the instructor may wish to expand upon the applications in Section 3.6. If time permits, the instructor may choose to require the students to complete a project, perhaps along the lines of one of the applications in that section. The book could also serve as a supplemental text for a one- or two-semester course in digital communications or signal processing.

This book is designed for self-study by engineers and beginning graduate students, and the manuscript has been used in that way by many readers over the past several years. The basic tools of linear system analysis are reviewed and integrated into the exposition of second-order random processes. The discussions are sufficiently detailed to walk the reader through the applications of the concepts and techniques that are presented. Several examples and exercises with solutions are provided to test the reader's understanding along the way. Each chapter has a set of problems that further test the reader's understanding and extend some of the topics presented in the text.

I wish to thank each of the instructors who taught from the manuscript for the book and supplied suggestions and corrections. Special thanks are due Professor Dilip Sarwate of the University of Illinois, Professor James Lehnert of Purdue University, and Professors John Komo and Daniel Noneaker of Clemson University. Each was kind enough to teach from one or more versions of the manuscript and provide extensive feedback that improved the book. Finally, I wish to express my appreciation to the students who suffered through numerous revisions of the manuscript and furnished lists of corrections.

MICHAEL B. PURSLEY

Introduction

The measurement of the resistance of a length of wire is subject to error. Assuming that the environment (e.g., the ambient temperature) around the wire is constant, we can model the measured resistance as a random variable. If the ambient environment fluctuates randomly, so does the resistance of the wire. In this scenario, there are two sources of randomness: the measurement error and the environment. The fluctuating resistance cannot be modeled as a single random variable; instead, it is a time-varying random phenomenon. We might measure the resistance at specific times, in which case we could model the results by using one random variable for each measurement. Such a random variable would incorporate the effects of both sources of randomness. Suppose the measurements are repeated once every 30 minutes, until 10 measurements are made. In this situation, we need only a finite number of random variables to model the measurements of the resistance of the wire.

If the ambient conditions vary continuously, the resistance of the wire also varies continuously. If we want a detailed mathematical description of how the measured resistance behaves as a function of time, it may not be sufficient to use only a finite collection of random variables. However, we can view the fluctuating resistance as a waveform that varies with time in a random manner. Alternatively, we can model the measured resistance of the wire at each instant of time as a random variable, thereby requiring an infinite number of random variables to model the resistance over the time interval of interest. If there is noise in the system in which the wire is used, the current in the wire may fluctuate randomly, even if the resistance is constant. To model this time-varying current in a way that reflects its randomness, we again must depart from the use of a single random variable or even a finite number of random variables.

In this book, the concept of a random variable is extended to the notion of a random process. For the time being, think of a random process as one of two entities: a random waveform or an infinite collection of random variables. Each of these two points of view is explored in detail in Chapter 2, but for now the reader may employ whichever conceptual model seems more natural in the context of the example under consideration.

A wide variety of engineering problems simply cannot be formulated in terms of a finite number of random variables. Just as a continuous-time signal at a particular point in an electronic system cannot be specified by finitely many real numbers, the noise in that system cannot be described by a finite number of real random variables. We can, however, model the noise voltage at the point of interest as an infinite collection of random variables: Let X_t be the random variable that specifies the voltage at time t for $-\infty < t < \infty$. Even if we are interested in the voltage for only a finite interval of time, such as $0 \leq t \leq T$, there are infinitely many instants in this interval, and we therefore require infinitely many random variables. One can also think of this noise as a random waveform. For example, it may be possible to start a strip-chart recorder at time $t = 0$ and record the noise voltage for T seconds. The recorded graph represents the noise voltage for the interval $0 \leq t \leq T$. Many engineering problems such as these require the use of one or more random processes to model the random phenomena of interest. Some engineering problems could conceivably be handled with a finite number of random variables, but the number is so large that it is actually easier to carry out the analysis if we use a random process instead.

To answer certain types of questions that arise in engineering applications, engineers must go beyond the notion of a random variable or a finite collection of random variables in order to formulate the relevant mathematical problems and find their solutions. For example, consider the problem of predicting future values of noise in a radar receiver on the basis of past values of the noise, or the problem of designing a filter to smooth or otherwise reduce the deleterious effects of the noise in an audio system. Since the noise process is observable at any instant of time, one has to deal with an infinite number of random variables in the solution of such problems. Other examples of phenomena that are best described as random processes are the forces acting on a high-performance aircraft during a difficult maneuver and the fluctuations in the temperature at a point on the earth. Such phenomena are virtually impossible to describe in a deterministic fashion, and they are not readily characterized by a few random variables.

The purpose of this book is to introduce the concept of a random process, examine certain problems that deal with the linear filtering of random processes, and develop techniques for the solutions of such problems. The book bridges the gap between the typical book or course in probability and the typical book or course in digital communications or stochastic signal processing. A firm understanding of second-order random processes is essential for work in such areas as wireless communications, and the book is designed to provide the requisite understanding.

The book is suitable as a text for advanced undergraduate students or beginning graduate students. It is also useful as a supplemental text in undergraduate courses in probability, undergraduate courses in analog or digital communications, and graduate courses in digital communications. Practicing engineers and beginning graduate students will find the book appropriate for self-study. All of the details are provided in the development of the principal tools for the analysis of random processes with applications in communication, control, and signal-processing systems. Several examples and exercises with solutions are provided to test the reader's understanding throughout the development of the material, and additional exercises appear at the end of each chapter.

It is assumed that the reader is familiar with basic probability theory, random variables, distribution functions, and density functions for discrete and continuous random variables. The very brief review of basic probability theory and random variables given in Chapter 1 is intended primarily to establish some notation and terminology that are used in the subsequent chapters. The material presented in the book focuses on the second-order analysis of real random processes, so it deals primarily with correlation functions and spectral densities. The basic definitions and descriptions for random processes and their correlation functions are given in Chapter 2. In Chapter 3, methods are developed for the analysis and design of linear filters that have random processes as their inputs. Chapter 4 is devoted to the concept of the spectral density function and its use in the analysis of random processes in linear systems and in the modeling of communications signals. The presentations in Chapters 3 and 4 require the reader to understand basic linear system analysis, including the use of convolutions and Fourier transforms. As a result a previous course in linear system analysis is a prerequisite for this book.

A few of the equations are so closely linked that it is convenient to refer to them as a set. Such equations may be assigned the same number, but they are distinguished by different letters. The set of equations is then referred to by number only. For example, a reference to (4.50) should be interpreted as a reference to the pair of equations numbered (4.50a) and (4.50b).

C H A P T E R   1

Probability and Random Variables: Review and Notation

1.0 PURPOSE OF THE CHAPTER

The purpose of this chapter is to review the essential facts from the theory of probability and random variables that are necessary for an understanding of the material on random processes in Chapters 2 through 4 and for subsequent use in the analysis of digital communication systems. Because it is assumed that the reader has taken a previous course in probability and random variables, the present chapter is considerably more terse than those that follow, and no claim is made regarding the completeness of the review presented. The reader may refer to any of the several excellent texts on probability and random variables in the literature, such as those listed at the end of the chapter, for more complete discussions of the topics presented and for examples and exercises to illustrate the material.

1.1 PROBABILITY SPACES

A *probability space* is a triple $(\Omega, \mathcal{F}, P)$ consisting of a set Ω, a collection $\mathcal{F}$ of subsets of Ω, and a function P defined on $\mathcal{F}$. The set Ω is the *sample space*, $\mathcal{F}$ is the *event class*, and P is the *probability measure*. The elements of the collection $\mathcal{F}$ are the *events*, which are the subsets of Ω to which we wish to assign probabilities. In order to be an event class, a collection $\mathcal{F}$ of subsets of Ω must satisfy certain conditions so that we can assign probabilities in a consistent way to all the events of interest. To say that the set A is an event is equivalent to saying that A is an element of $\mathcal{F}$, and this is denoted by $A \in \mathcal{F}$. The basic properties of an *event class* $\mathcal{F}$ are (1) $\Omega \in \mathcal{F}$, (2) if $A \in \mathcal{F}$, then so is its complement A^c, and (3) if each of the sets in the countable collection $A_1, A_2, A_3, \ldots$ is in $\mathcal{F}$, so is the union of all of the sets in this collection. A *countable* collection or set is a set whose elements can be counted; that is, the elements can be put in a one-to-one correspondence with a subset of the positive integers. Such a collection or set is often referred to as a *discrete* collection or set. Of course, *any finite set is countable*, but many *infinite* sets are also countable, such as the set of all integers (positive and negative), the set of all fractions of the form $1/n$, where n is a positive integer, and the set of all rational numbers. An example of a set that is *not* countable is the set of all real numbers.

To summarize, the event class $\mathcal{F}$ (1) includes the sample space Ω (Ω is an event), (2) is closed under complementation (the complement of an event is an event), and

(3) is closed under countable unions (the union of a countable collection of events is an event). It can be shown that if $\mathcal{F}$ is an event class, then it is closed under countable *intersections* as well; that is, countable intersections of events are also events.

The probability measure P assigns probabilities to events in a consistent way. If A is an event, then $P(A)$, the *probability of the event A*, is a nonnegative number that does not exceed unity (i.e., $0 \le P(A) \le 1$ for each $A \in \mathcal{F}$). The probability of the entire sample space is unity (i.e., $P(\Omega) = 1$). If $A_1, A_2, A_3, \ldots$ is a countable collection of *disjoint* events, the probability of the union of all of these events is the sum of the probabilities: $P(A_1 \cup A_2 \cup A_3 \cup \ldots) = P(A_1) + P(A_2) + P(A_3) + \ldots$. Recall that events A and B are *disjoint* if their intersection $A \cap B$ is empty. Disjoint events are also referred to as being *mutually exclusive*.

1.2   RANDOM VARIABLES

For the time being, it is sufficient to consider *real* random variables and vectors whose components are such random variables. In an engineering problem, a random variable may model the number of phone calls received during a particular interval, the voltage at a certain point in a noisy electronic system, the number of bits in error in a word stored in a computer memory, or the number of failures of a certain type of electronic component. In a communication receiver, random variables arise in many ways. For example, a random variable results if the output of the receiver's filter is sampled at a single instant. The randomness in this output is due in part to the thermal noise in the receiver. If the filter output is sampled several times, the resulting samples can be grouped together to form a vector whose components are random variables. For some applications, it is convenient to consider complex random variables, but such random variables are not needed in the present chapter. In this book, unless it is stated otherwise, a random variable is assumed to be a real random variable.

The formal definition of a real random variable X on a probability space $(\Omega, \mathcal{F}, P)$ is obtained by viewing X as a real-valued function defined on the sample space Ω: For each point ω in the sample space, $X(\omega)$ is a real number. Not just any such function will do; for mathematical and physical reasons, in order for X to be a random variable, it must be that

$$\{\omega \in \Omega : X(\omega) \le u\} \in \mathcal{F}$$

for each real number u. In words, this says that, for an arbitrary real number u, the set of all points ω for which $X(\omega) \le u$ must be an event. The set $S = \{\omega \in \Omega : X(\omega) \le u\}$ is illustrated in Figure 1–1. Notice that the point ω is in the set S, while the point ω' is not.

The reason we insist that $S = \{\omega \in \Omega : X(\omega) \le u\}$ be an event for each u is that, for a random variable to be of any value in engineering applications, we need to be able to inquire about the probability that the random variable does not exceed some number. For example, we may be interested in the probability that the number of failures does not exceed the design limit for a fault-tolerant system, the probability that the voltage does not exceed a certain threshold in an electronic system, or the probability that the number of incoming phone calls does not exceed the number of operators at the switchboard. In order to be able even to pose the problem of finding the probability that the

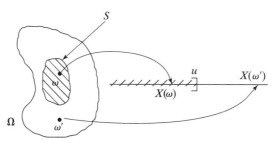

FIGURE 1–1   The event $S = \{\omega \in \Omega : X(\omega) \leq u\}$.

random variable X does not exceed u, it is necessary for S to be an event so it can be assigned a probability.

The probability that we seek in the preceding paragraph is $P(\{\omega \in \Omega : X(\omega) \leq u\})$. The notation $\{X \leq u\}$ is shorthand for $\{\omega \in \Omega : X(\omega) \leq u\}$, and $P(X \leq u)$ is shorthand notation for the probability $P(\{\omega \in \Omega : X(\omega) \leq u\})$. Such probabilities can be defined and their values sought if X is a random variable.

If $X_1, X_2, X_3, \ldots, X_n$ are all random variables on the same probability space $(\Omega, \mathscr{F}, P)$, then $\{X_k \leq u_k\}$ is an event for each k in the range $1 \leq k \leq n$ and each choice of the real numbers $u_1, u_2, \ldots, u_n$. It follows that

$$\bigcap_{k=1}^{n} \{X_k \leq u_k\} \in \mathscr{F},$$

so probabilities of the form

$$P(X_1 \leq u_1, X_2 \leq u_2, \ldots, X_n \leq u_n) = P\left(\bigcap_{k=1}^{n} \{X_k \leq u_k\}\right)$$

are defined for each positive integer n and each choice of $u_1, u_2, \ldots, u_n$. It is often useful to consider the n random variables $X_1, X_2, \ldots, X_n$ as an n-dimensional *random vector* $\mathbf{X} = (X_1, X_2, \ldots, X_n)$. Each component of $\mathbf{X}$ is a real random variable.

In engineering we are concerned with the application of probabilistic methods to the solution of practical problems. Consequently, we usually need not concern ourselves with the detailed mathematical structure of the probability space or the random variable. In particular, we can generally assume that the random phenomenon encountered in practice can be modeled in terms of valid random variables for *some* choice of the probability space $(\Omega, \mathscr{F}, P)$, so we study the detailed structure of $X(\omega)$ only if it is convenient to do so and only when it is helpful in solving the problem at hand. We are usually able to derive all of the necessary information from the distribution or density functions, the subject of the next section.

1.2.1   Distribution and Density Functions for Random Variables

The *distribution function*, also referred to as the cumulative distribution function, for a random variable X is denoted by F_X and defined by

$$F_X(u) = P(X \leq u) = P(\{\omega : X(\omega) \leq u\})$$

for each real number u. For each positive integer n, the distribution function for the random vector $\mathbf{X} = (X_1, X_2, \ldots, X_n)$ is the n-dimensional distribution function denoted by $F_{\mathbf{X},n}$ and defined by

$$F_{\mathbf{X},n}(u_1, u_2, \ldots, u_n) = P(X_1 \leq u_1, X_2 \leq u_2, \ldots, X_n \leq u_n),$$

which is equivalent to

$$
\begin{aligned}
F_{\mathbf{X},n}(u_1, \ldots, u_n) &= P\left[\bigcap_{k=1}^{n} \{X_k \leq u_k\} \right] \\
&= P\left[\bigcap_{k=1}^{n} \{\omega \in \Omega : X_k(\omega) \leq u_k\} \right]
\end{aligned}
\tag{1.1}
$$

for each choice of the real numbers u_k ($1 \leq k \leq n$). As discussed in Section 1.1, the fact that each X_k is a random variable guarantees that the intersection in (1.1) is an event.

The rationale for the notation $F_{\mathbf{X},n}$ is as follows: The first subscript identifies the random vector in question, and the second subscript gives the length of that random vector, which is also the dimension of the distribution. This notation is particularly useful in considering samples of a random process, as in Chapters 2 and 3. In such applications, the number of samples may vary throughout the discussion.

The concept of the distribution function for a random variable is a special case ($n = 1$) of the concept of a distribution function for a random vector. Rather than denote this as $F_{\mathbf{X},1}$, we usually just write F_X as the notation for the distribution function for a one-dimensional random vector (i.e., a one-dimensional random vector $\mathbf{X}$ whose only component is the random variable X). If $\mathbf{X} = (X_1, X_2)$, an alternative notation for $F_{\mathbf{X},2}(u_1, u_2)$ is $F_{X_1, X_2}(u_1, u_2)$, and the latter is usually referred to as the joint distribution function for the random variables X_1 and X_2.

The two random variables X_1 and X_2 are *statistically independent* if and only if their joint distribution function factors—that is, if and only if $F_{X_1, X_2}(u_1, u_2) = F_{X_1}(u_1)F_{X_2}(u_2)$ for all choices of u_1 and u_2. The functions F_{X_1} and F_{X_2} are the *marginal distribution functions* for the random variables X_1 and X_2, respectively. The random variables $X_1, X_2, \ldots, X_n$ are *mutually independent* if and only if the n-dimensional distribution function $F_{\mathbf{X},n}(u_1, u_2, \ldots, u_n)$ factors as the product of the marginal distribution functions of the individual random variables.

If the distribution function $F_{\mathbf{X},n}$ has an associated *density* function $f_{\mathbf{X},n}$, then we say that the random vector $\mathbf{X} = (X_1, X_2, \ldots, X_n)$ is a *continuous* random vector with density $f_{\mathbf{X},n}$. This means that

$$F_{\mathbf{X},n}(u_1, \ldots, u_n) = \int_{-\infty}^{u_n} \cdots \int_{-\infty}^{u_1} f_{\mathbf{X},n}(y_1, \ldots, y_n)\, dy_1 \ldots dy_n \tag{1.2a}$$

for each n and each choice of $\mathbf{u} = (u_1, \ldots, u_n)$. It follows that we can let

$$f_{\mathbf{X},n}(y_1, \ldots, y_n) = \partial^n F_{\mathbf{X},n}(u_1, \ldots, u_n)/\partial u_1 \ldots \partial u_n|_{\mathbf{u}=\mathbf{y}} \tag{1.2b}$$

for all $\mathbf{y} = (y_1, \ldots, y_n)$ for which the derivative exists. As with distribution functions, we usually drop the subscript "1" in the notation for the (one-dimensional) density function of a random variable. Thus, if X is a continuous random variable having distribution function F_X, its density f_X satisfies the relation

$$F_X(u) = \int_{-\infty}^{u} f_X(y)\, dy \tag{1.3a}$$

for each choice of u. It follows that we can let

$$f_X(y) = dF_X(u)/du|_{u=y} \tag{1.3b}$$

for all values of y for which the derivative exists.

Next, we consider a certain class of random variables that do not have continuous distribution functions. Let $\mathbb{Z}$ be the set of all integers (positive, negative, and zero), and let $\mathbb{I}$ be any subset of $\mathbb{Z}$. Consider random variables that take on values in a discrete ordered set $S = \{s_k : k \in \mathbb{I}\}$. In this book, by *ordered* we mean that $s_i < s_j$ whenever $i < j$. We can assume that $\{X(\omega) : \omega \in \Omega\} = S$, but it usually suffices to require only the slightly weaker condition that $P(X \in S) = 1$. Two important examples of discrete sets that fit our definition of *ordered* are $S = \mathbb{Z}$ and $S = \{0, 1, 2, \ldots\}$, the set of nonnegative integers. A random variable of this type is called a *discrete random variable*. For a discrete random variable, we can always define a *discrete density function* (also known as a probability mass function) from its distribution function. The discrete density function for such a random variable X is given by

$$f_X(s_i) = F_X(s_i) - F_X(s_{i-1}), \tag{1.4a}$$

which is just $P(X = s_i)$, the probability that X takes on the value s_i. The expression on the right-hand side is the difference in the value of the distribution function at s_i and the next smaller value in S. On the other hand, given the discrete density function f_X, the corresponding distribution function is given by

$$F_X(s_i) = \sum_{k \le i} f_X(s_k). \tag{1.4b}$$

If $S = \mathbb{Z}$, the sum in (1.4b) can be written as

$$F_X(s_i) = \sum_{k=-\infty}^{i} f_X(s_k),$$

and if $S = \{0, 1, 2, \ldots\}$, the sum is equivalent to

$$F_X(s_i) = \sum_{k=0}^{i} f_X(s_k).$$

A typical example of a discrete random variable is the number of successes in a sequence of independent trials of an experiment that can have only two outcomes. The simplest illustration is a sequence of n tosses of a biased coin that comes up "heads" with probability p and "tails" with probability $1 - p$. The probability of the occurrence of a *particular sequence* of k heads and $(n - k)$ tails is $p^k(1 - p)^{n-k}$. If X denotes the random variable representing the number of heads out of n independent tosses of the coin, then

$$f_X(k) = \binom{n}{k} p^k (1 - p)^{n-k}$$

for $0 \leq k \leq n$. This is just a result of multiplying the number of sequences with exactly k heads by the probability of each such sequence. A random variable X with this discrete density function is said to have a *binomial distribution* with parameters n and p.

If a two-dimensional random vector $\mathbf{X} = (X_1, X_2)$ is such that both X_1 and X_2 take values in the same discrete ordered set, the discrete density is

$$f_{\mathbf{X},2}(s_i, s_j) = F_{\mathbf{X},2}(s_i, s_j) - F_{\mathbf{X},2}(s_{i-1}, s_j) - F_{\mathbf{X},2}(s_i, s_{j-1}) + F_{\mathbf{X},2}(s_{i-1}, s_{j-1}). \quad (1.5a)$$

On the other hand, given the two-dimensional discrete density function $f_{\mathbf{X},2}$, the corresponding distribution function is

$$F_{\mathbf{X},2}(s_i, s_j) = \sum_{k \leq i} \sum_{m \leq j} f_{\mathbf{X},2}(s_k, s_m). \quad (1.5b)$$

Similar equations can be written for general n-dimensional distribution and density functions.

It is clear that a random variable cannot have *both* a density function [as in (1.3)] and a discrete density function [as in (1.4)]. Thus, in referring to the *density of a random variable,* we mean the form of density that is appropriate for the particular random variable in question. Obviously, for a discrete random variable, only a discrete density function is appropriate; for a continuous random variable, the only density that can be considered is a density of the form given in (1.3).

1.2.2 Means, Variances, and Moments

If X is a random variable and g is a well-behaved function, then $Y = g(X)$ is also a random variable. If X is a continuous random variable, the expected value of $g(X)$ can be evaluated from the expression

$$E\{g(X)\} = \int_{-\infty}^{\infty} g(u) f_X(u) \, du, \quad (1.6)$$

provided that the integral exists. If X is a discrete random variable, $E\{g(X)\}$ is given by a similar expression in which the integral is replaced by a sum. The mean, variance, and moments of the continuous random variable X can all be expressed as special cases of (1.6). The *mean* of X is $E\{X\}$ which is just (1.6) with $g(u) = u$ for all u; the *second moment* of X is $E\{X^2\}$, which corresponds to $g(u) = u^2$ for all u; and the mth moment for any positive integer m is $E\{X^m\}$, which corresponds to $g(u) = u^m$. If μ denotes the mean of X, the variance of X is $\mathrm{Var}\{X\} = E\{(X - \mu)^2\}$, and this corresponds to letting $g(u) = (u - \mu)^2$ for all u. The standard deviation of X, usually denoted by σ or σ_X, is just $\sqrt{\mathrm{Var}\{X\}}$.

If X_1 and X_2 are random variables and g is a well-behaved function, then $Y = g(X_1, X_2)$ is also a random variable. The mean of Y is

$$E\{g(X_1, X_2)\} = \int_{-\infty}^{\infty} \int_{-\infty}^{\infty} g(u_1, u_2) f_{X_1, X_2}(u_1, u_2) \, du_1 \, du_2, \quad (1.7)$$

provided that the integral exists. The covariance and correlation for the pair X_1 and X_2 can each be expressed as a special case of (1.7). If $g(u_1, u_2) = u_1 u_2$, then (1.7) gives the defining expression for the *correlation* $E\{X_1 X_2\}$; if $g(u_1, u_2) = (u_1 - \mu_1)(u_2 - \mu_2)$, where $\mu_1 = E\{X_1\}$ and $\mu_2 = E\{X_2\}$, then (1.7) gives the definition of the covariance

$$\text{Cov}\{X_1, X_2\} = E\{(X_1 - \mu_1)(X_2 - \mu_2)\};$$

and if $g(u_1, u_2) = (u_1 - \mu_1)(u_2 - \mu_2)/\sigma_1\sigma_2$, where $\sigma_1 = \sqrt{\text{Var}\{X_1\}}$ and $\sigma_2 = \sqrt{\text{Var}\{X_2\}}$, then (1.7) gives the definition of the correlation coefficient, often denoted by ρ.

1.2.3   Gaussian Random Variables

A *Gaussian random variable* X is a continuous random variable with density function of the form

$$f_X(u) = \frac{\exp\{-(u - \mu)^2/2\sigma^2\}}{\sqrt{2\pi\sigma^2}}, \tag{1.8}$$

where μ is the mean of the random variable X and σ^2 is the variance of X. The corresponding distribution function cannot be written in closed form, but it can be expressed in terms of the tabulated integral

$$\Phi(x) = \int_{-\infty}^{x} \frac{\exp(-y^2/2)}{\sqrt{2\pi}} \, dy, -\infty < x < \infty. \tag{1.9}$$

If X is a Gaussian random variable with mean μ and variance σ^2, then

$$F_X(u) = \Phi((u - \mu)/\sigma), \quad -\infty < u < \infty.$$

Error probabilities for digital communication systems are best expressed in terms of the complementary distribution function, which is defined by

$$Q(x) = \int_{x}^{\infty} \frac{\exp(-y^2/2)}{\sqrt{2\pi}} \, dy, -\infty < x < \infty. \tag{1.10}$$

It should be clear from (1.9) and (1.10) that $Q(x) + \Phi(x) = 1$. Also, replacing x by $-x$ in (1.9), followed by the change of variable $u = -y$, shows that

$$\Phi(-x) = \int_{x}^{\infty} \frac{\exp(-u^2/2)}{\sqrt{2\pi}} \, du = Q(x).$$

It follows that the relationship between the two functions is $Q(x) = 1 - \Phi(x) = \Phi(-x)$ for all x in the range $-\infty < x < \infty$.

Although there is no exact closed-form expression for the function Q, several good bounds and approximations can be computed easily. In [1.2], it is shown that the family of functions

$$G_{a,b}(x) = \left\{(1 - a)x + a\sqrt{x^2 + b}\right\}^{-1} \exp(-x^2/2)/\sqrt{2\pi} \tag{1.11}$$

is very useful in obtaining bounds and approximations for $Q(x)$ that are appropriate for performance analyses of digital communication systems. In particular, if $a = 1/\pi$ and $b = 2\pi$, a lower bound on $Q(x)$ is obtained. By substituting for a and b in (1.11), it is easy to show that this lower bound is given by

$$Q_1(x) = \sqrt{\pi/2}\left\{(\pi - 1)x + \sqrt{x^2 + 2\pi}\right\}^{-1} \exp(-x^2/2). \tag{1.12}$$

An upper bound, which we denote by $Q_2(x)$, is obtained if the values of a and b in (1.11) are $a = 0.344$ and $b = 5.334$. It is also pointed out in [1.2] that a good approximation, which we denote by $Q_3(x)$, results from (1.11) by letting $a = 0.339$ and $b = 5.510$. Thus, we have $Q_1(x) \leq Q(x) \leq Q_2(x)$ and $Q(x) \approx Q_3(x)$ for values of x that are of interest in the evaluation of the performance of digital communication systems. See [1.2] for further discussion and additional references on this family of bounds and approximations.

A family of series approximations for the function Q is given in [1.1]. For one choice of parameters, this approximation can be written as

$$\widetilde{Q}(x) = 0.5 - \left(\frac{2}{\pi}\right) H(x), \tag{1.13a}$$

where

$$H(x) = \sum_{\substack{n=1 \\ n \text{ odd}}}^{33} n^{-1} \exp\{-(n\pi)^2/392\} \sin(n\pi x/14). \tag{1.13b}$$

There are many situations in engineering in which two or more random variables are Gaussian, not only individually, but also collectively. This concept is made precise by the consideration of the *joint* distribution for the random variables. One way to define jointly Gaussian random variables is to give their joint density function: The random variables X and Y are said to be *jointly Gaussian* if their joint density function is of the form

$$f_{X,Y}(u, v) = \frac{\exp\left\{\dfrac{-1}{2(1 - \rho^2)}\left[\left(\dfrac{u - \mu_1}{\sigma_1}\right)^2 - 2\rho\left(\dfrac{u - \mu_1}{\sigma_1}\right)\left(\dfrac{v - \mu_2}{\sigma_2}\right) + \left(\dfrac{v - \mu_2}{\sigma_2}\right)^2\right]\right\}}{2\pi\sigma_1\sigma_2\sqrt{(1 - \rho^2)}},$$

where μ_1 and σ_1 are the mean and standard deviation for X, μ_2 and σ_2 are the mean and standard deviation for Y, and

$$\rho = \frac{E\{(X - \mu_1)(Y - \mu_2)\}}{\sigma_1\sigma_2} = \frac{\text{Cov}\{X, Y\}}{\sigma_1\sigma_2}$$

is the correlation coefficient. An important feature of jointly Gaussian random variables is that their joint density function is completely determined by the five parameters $\mu_1, \mu_2, \sigma_1, \sigma_2$, and ρ. Also, it is easy to show that jointly Gaussian random variables are independent if and only if their correlation coefficient is zero ($\rho = 0$).

For generalization to more than two random variables, it is more convenient to define a pair of random variables X and Y to be jointly Gaussian if each linear combination of X and Y is a Gaussian random variable; that is, X and Y are jointly Gaussian if $\alpha X + \beta Y$ is a Gaussian random variable for each choice of the real numbers α and β. It can be shown that this definition is equivalent to the condition that X and Y have the joint density function $f_{X,Y}$ given in the previous paragraph. (See pp. 46–49 of [1.9] or pp. 156 and 210 of [1.10].)

Consider a set of n random variables $X_1, X_2, \ldots, X_n$. We may treat these collectively by defining the random vector $\mathbf{X} = (X_1, X_2, \ldots, X_n)$, and then the density function for the random vector $\mathbf{X}$ is just the joint density function for the random variables $X_1, X_2, \ldots, X_n$. We say that a set of n random variables $X_1, X_2, \ldots, X_n$ is jointly Gaussian if every linear combination of these random variables is a Gaussian random variable. It follows from this definition that the n random variables are jointly Gaussian if and only if they have a joint density function that is of the Gaussian form. Specifically, if Λ is the $n \times n$ matrix with

$$\Lambda_{i,j} = E\{(X_i - \mu_i)(X_j - \mu_j)\}$$

as the element in the ith row and jth column, and if μ_i is the mean of X_i $(1 \le i \le n)$, then the joint density function for $\mathbf{X}$ is given by

$$f_{\mathbf{X},n}(u) = (2\pi)^{-n/2}|\det(\Lambda)|^{-1/2}\exp\left\{-\tfrac{1}{2}(\mathbf{u} - \boldsymbol{\mu})\Lambda^{-1}(\mathbf{u} - \boldsymbol{\mu})^T\right\},$$

where $\det(\Lambda)$ is the determinant of the matrix Λ, Λ^{-1} is the inverse of the matrix Λ, $\mathbf{u} = (u_1, u_2, \ldots, u_n)$, $\boldsymbol{\mu} = (\mu_1, \mu_2, \ldots, \mu_n)$, and $(\mathbf{u} - \boldsymbol{\mu})^T$ is the transpose of the vector $(\mathbf{u} - \boldsymbol{\mu})$. The matrix Λ is known as the covariance matrix for the random vector $\mathbf{X}$, and the vector $\boldsymbol{\mu}$ is the mean vector for $\mathbf{X}$. For $n = 2$, this n-dimensional density function reduces to the two-dimensional density

$$f_{X,Y}(u,v) = \frac{\exp\left\{\dfrac{-1}{2(1-\rho^2)}\left[\left(\dfrac{u-\mu_1}{\sigma_1}\right)^2 - 2\rho\left(\dfrac{u-\mu_1}{\sigma_1}\right)\left(\dfrac{v-\mu_2}{\sigma_2}\right) + \left(\dfrac{v-\mu_2}{\sigma_2}\right)^2\right]\right\}}{2\pi\sigma_1\sigma_2\sqrt{(1-\rho^2)}}$$

if we set $X_1 = X$, $X_2 = Y$, $\Lambda_{1,1} = \sigma_1^2$, $\Lambda_{2,2} = \sigma_2^2$, and $\Lambda_{2,1} = \Lambda_{1,2} = \rho\sigma_1\sigma_2$.

Several important problems that arise in digital communications deal with a pair of independent Gaussian random variables. If X and Y are jointly Gaussian and independent, they have the joint density function just given, with $\rho = 0$. But if $\rho = 0$, this density is

$$f_{X,Y}(u,v) = \frac{\exp\{-(u-\mu_1)^2/2\sigma_1^2\}\exp\{-(v-\mu_2)^2/2\sigma_2^2\}}{2\pi\sigma_1\sigma_2}. \tag{1.14}$$

Of course, X and Y are individually Gaussian, so they have densities

$$f_X(u) = \exp\{-(u-\mu_1)^2/2\sigma_1^2\}/\sqrt{2\pi}\sigma_1 \tag{1.15}$$

and

$$f_Y(v) = \exp\{-(v-\mu_2)^2/2\sigma_2^2\}/\sqrt{2\pi}\sigma_2 \tag{1.16}$$

Notice that (1.14)–(1.16) show that $\rho = 0$ implies $f_{X,Y}(u,v) = f_X(u)f_Y(v)$ for each value of u and v, which establishes that uncorrelatedness implies independence for a pair of jointly Gaussian random variables.

The problem that we wish to address is the determination of the probability that the random pair (X, Y) falls in a rectangular region of the form

$$S = \{(u,v) : x_1 \le u \le x_2, y_1 \le v \le y_2\},$$

for an arbitrary choice of x_1, x_2, y_1, and y_2, subject only to the constraint that $x_1 < x_2$ and $y_1 < y_2$. This set is illustrated in Figure 1–2.

The easiest way to evaluate $P((X,Y) \in S)$ is to observe that

$$P((X,Y) \in S) = P(x_1 \le X \le x_2, y_1 \le Y \le y_2)$$

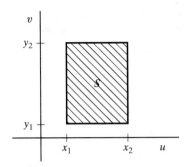

FIGURE 1–2   The region S.

and, because X and Y are independent,

$$P(x_1 \leq X \leq x_2, y_1 \leq Y \leq y_2) = P(x_1 \leq X \leq x_2)P(y_1 \leq Y \leq y_2) =$$
$$[F_X(x_2) - F_X(x_1)][F_Y(y_2) - F_Y(y_1)].$$

Since X and Y are Gaussian, it follows that

$$F_X(x) = \Phi((x - \mu_1)/\sigma_1)$$

for each x and

$$F_Y(y) = \Phi((y - \mu_2)/\sigma_2)$$

for each y. Consequently,

$$P((X, Y) \in S) =$$
$$[\Phi((x_2 - \mu_1)/\sigma_1) - \Phi((x_1 - \mu_1)/\sigma_1)][\Phi((y_2 - \mu_2)/\sigma_2) - \Phi((y_1 - \mu_2)/\sigma_2)] \quad (1.17)$$

In (1.17), it is permissible to let x_2 or y_2 be $+\infty$ and x_1 or y_1 be $-\infty$. For example, if $y_2 = +\infty$ and $x_2 = +\infty$, then

$$\Phi((x_2 - \mu_1)/\sigma_1) = 1$$

and

$$\Phi((y_2 - \mu_2)/\sigma_2) = 1,$$

so that

$$P((X, Y) \in S) = [1 - \Phi((x_1 - \mu_1)/\sigma_1)][1 - \Phi((y_1 - \mu_2)/\sigma_2)]$$
$$= Q((x_1 - \mu_1)/\sigma_1)Q((y_1 - \mu_2)/\sigma_2). \quad (1.18)$$

REFERENCES AND SUGGESTIONS FOR FURTHER READING

[1.1]   N. C. Beaulieu, "A Simple Series for Personal Computer Computation of the Error Function $Q(\cdot)$," *IEEE Transactions on Communications*, vol. 37, September 1989, pp. 989–991.

[1.2]   P. O. Börjesson and C.-E. W. Sundberg, "Simple Approximations of the Error Function $Q(x)$ for Communications Applications," *IEEE Transactions on Communications*, vol. COM-27, March 1979, pp. 639–643.

[1.3] G. R. Cooper and C. D. McGillem, *Probabilistic Methods of Signal and System Analysis* (2nd ed.). New York: Holt, Rinehart, and Winston, 1986.

[1.4] C. W. Helstrom, *Probability and Stochastic Processes for Engineers* (2nd ed.). New York: Macmillan, 1991.

[1.5] P. G. Hoel, S. C. Port, and C. J. Stone, *Introduction to Probability Theory*. Boston: Houghton Mifflin, 1971.

[1.6] J. J. Komo, *Random Signal Analysis in Engineering Systems*. New York: Academic Press, 1987.

[1.7] A. Leon-Garcia, *Probability and Random Processes for Electrical Engineering*. Reading, MA: Addison-Wesley, 1989.

[1.8] S. Ross, *A First Course in Probability* (6th ed.). New York: Macmillan, 2002.

[1.9] E. Wong and B. Hajek, *Stochastic Processes in Engineering Systems*. New York: Springer-Verlag, 1985.

[1.10] J. M. Wozencraft and I. M. Jacobs, *Principles of Communication Engineering*. New York: Wiley, 1965.

PROBLEMS

1.1 A sequence of binary digits is transmitted in a certain communication system. Any given digit is received erroneously with probability p and received correctly with probability $1 - p$. Errors occur independently from digit to digit. Out of a sequence of n digits transmitted, what is the probability that no more than j digits are received erroneously?

1.2 A certain error-correcting code is applied to the communication system described in Problem 1.1.
  (a) The code can correct a single error in any digit, but it cannot correct any pattern of two or more errors. What is the probability that the code can correct the error pattern that occurs when n digits are transmitted?
  (b) The code can detect any pattern of two or fewer errors. What is the probability that the code can detect the error pattern that occurs when n digits are transmitted?

1.3 Two random variables X_1 and X_2 are produced in a digital communication receiver. An error occurs if $X_1 > X_2$; otherwise, the receiver output is correct. The random variables X_1 and X_2 are independent Gaussian random variables with means μ_1 and μ_2 and standard deviations σ_1 and σ_2, respectively. Find the probability of error for this receiver. Express your answer in terms of the function $\Phi(\cdot)$, and then convert it to be in terms of the function $Q(\cdot)$. (*Hint*: Perhaps the best approach is to let $X = X_1 - X_2$, so that X is a Gaussian random variable and $P(X > 0)$ is the probability of error for the communication receiver.)

1.4 Suppose the random variables X_1, X_2, and X_3 are jointly Gaussian with means $\mu_i = E\{X_i\}$ and covariances $\Lambda_{i,j} = E\{(X_i - \mu_i)(X_j - \mu_j)\}, 1 \le i \le 3$ and $1 \le j \le 3$. Let Y be the random variable defined by $Y = X_1 + X_2 + X_3$. Express the distribution function for Y in terms of the function $\Phi(\cdot)$ and the parameters μ_i and $\Lambda_{i,j}$.

1.5 The decision statistic in a certain communication receiver is a Gaussian random variable Z with mean μ and standard deviation σ. The decision made by the receiver depends on the value of Z^2. Find the probability that $Z^2 < 2$. Note that Z^2 is *not* Gaussian!

1.6 Suppose that X_1 and X_2 are Gaussian random variables with means μ_1 and μ_2, respectively. Assume that $\mu_1 \ne \mu_2$. The variance for each of the two random variables is σ^2. Find the value of x for which $f_{X_1}(x) = f_{X_2}(x)$.

1.7 The random variables X_1 and X_2 are defined as in Problem 1.6. Show that, for each value of y, $f_{X_1}(y)$ is larger than $f_{X_2}(y)$ if and only if $|y - \mu_1|$ is smaller than $|y - \mu_2|$. This observation is employed in communication receivers that use maximum-likelihood statistical decisions.

1.8 The random variables X_1 and X_2 are Gaussian and independent with means μ_1 and μ_2, respectively. Each has variance σ^2. What is the probability that the point (X_1, X_2) is in the square with vertices $(1,1)$, $(1,2)$, $(2,1)$, and $(2,2)$? For what values of μ_1 and μ_2 is this probability maximized?

1.9 **(a)** Suppose that $x = \sqrt{2\alpha}$ and let α_{dB} be defined as $10 \log_{10}(\alpha)$. Evaluate $Q_1(x), Q_2(x)$, $Q_3(x)$, and $\widetilde{Q}(x)$ for $\alpha_{dB} = 10.5$.

(b) Evaluate the bounds in part **(a)** for values of α_{dB} ranging from 7.0 to 12.0 in steps of 0.1 dB. Give your answers by filling in the following table.

α_{dB}	$Q_1(\sqrt{2\alpha})$	$\widetilde{Q}(\sqrt{2\alpha})$	$Q_3(\sqrt{2\alpha})$	$Q_2(\sqrt{2\alpha})$
7.0				
7.1				
7.2				
.				
.				
.				
12.0				

1.10 It is often necessary to determine the value of α_{dB} for which $Q(\sqrt{2\alpha}) = 10^{-n}$ for some integer n. Use the bounds and approximations of Problem 1.9 to estimate, to the nearest tenth of a dB, the values of α_{dB} for which $Q(\sqrt{2\alpha}) = 10^{-n}$ for $n = 1, 2, \ldots, 8$.

Introduction to Random Processes

2.0   ORIGINS OF RANDOM PROCESSES IN ELECTRONIC SYSTEMS

Random processes are present in all electronic systems. Such a process may arise as a result of random motion of electrons in a resistive component of the system, or it may be due to some kind of human-made disturbance. Often, we model a signal generated by an electronic system as a random process, even though it may be deterministic in the sense that *if* we had an accurate model of every detail of the signal generation process and *if* we could solve all of the mathematical problems that arise in the analysis of that model, we could determine the values of the signal exactly at any given time. In practice, we rarely have such a model and we could not solve the analytical and computational problems required for a complete deterministic description of the signal. Moreover, for many applications, a deterministic characterization of the signal is unnecessary and would be far too complicated to be of any practical utility.

One example of a random process is the message process that is generated at the transmitter and conveys information to the receiver in a communication system such as a telephone system or a satellite broadcast link. The message process could be an analog video signal, a digitized speech signal, telemetry from a satellite or deep-space probe, or a sequence of binary digits from a digital computer. The message process may be modulated onto a radio-frequency (RF) signal (known as the carrier) that is suitable for transmission through the available communication medium. If so, the resulting RF signal represents a random process, not only because of the randomness in the message, but also because there may be a random phase associated with the unmodulated carrier itself.

In thinking of communication signals as represented by random processes, we may wish to take the point of view of the intended receiver, in which case the modeling of a signal as a random process reflects the fact that, before reception of the signal, the receiver does not know precisely the message that is being transmitted. If it did, there would be no need to transmit the signal at all! Sometimes it is simply the value of a parameter, such as the amplitude or phase of a sinusoidal signal, that is unknown to the intended receiver; if so, the randomness can be expressed via one or more random variables. In such a situation, modeling the signal as a random process may be more for convenience than necessity. Often, however, the unknown portion of the signal is more complex than this, and the randomness cannot be expressed in terms of a finite number of random variables. In such cases, modeling the signal as a random process is essential.

The various kinds of noise processes that arise in electrical and electronic devices are examples of highly complicated random processes that cannot be described in terms of finite numbers of random variables. Such noise processes may be due to automobile ignition, lightning, electromagnetic radiation from electronic equipment, radar signals from aircraft, or thermal noise from various components in a communication receiver. Strictly speaking, all electronic systems generate noise within the system and radiate noise to nearby systems. In order to be of any value, electronic equipment must be capable of satisfactory operation in the presence of such noise, so it is necessary to understand noise processes and their effects in order to be able to design efficient electronic systems.

In this chapter, basic mathematical models for random processes are introduced, and many properties of random processes that arise in engineering problems are presented. We begin with some examples.

2.1 EXAMPLES OF RANDOM PROCESSES

Before giving a precise definition of a random process, we present some of the basic intuitive notions that are needed in the study of random processes. This is accomplished by means of four examples that are typical of the random processes encountered in electrical and systems engineering. As a substitute for a precise definition of a *continuous-time random process*, consider such a process to be a collection of waveforms defined on a common interval (e.g., on $[0, \infty)$, (a, b), or $(-\infty, \infty)$). Randomness is introduced into this conceptual model by considering a hypothetical experiment in which one waveform is drawn at random from the collection of waveforms. Similarly, a *discrete-time random process* can be thought of as a collection of sequences, and the experiment is to draw one sequence at random from this collection.

Example 2–1   An Oscillator with a Random Phase

Consider an experiment in which an oscillator is switched on at some time T_0, and its output is observed during the time interval $0 \le t \le T$ (where $T_0 < 0 < T$). Assume that $T = f_0^{-1}$, where f_0 is the frequency of the oscillator, and let α be the amplitude of the sinusoidal signal at the output of the oscillator. The time T_0 is selected at random, but we assume that $|T_0|$ is sufficiently large that any transients have decayed.

Because T_0 is random (and typically, the oscillator's initial phase at time T_0 is also random), the sinusoidal signal at the output will have a random phase Θ at time $t = 0$. That is, the output will be

$$X(t) = \alpha \sin(2\pi f_0 t + \Theta)$$

for $0 \le t \le T$, as shown in Figure 2–1. The phase Θ is a random variable, and the nature of the random process $X(t)$ will depend on the distribution of Θ, which in turn depends on the distribution of T_0. We can consider the sinusoidal waveform $\alpha \sin(2\pi f_0 t + \Theta)$ for a particular value of the random variable Θ; such a waveform is called a *sample function* for the random process $X(t)$.

We can think of an experiment in which Θ is selected at random according to some distribution and the resulting sample function is sketched. If the experiment is repeated several times, we might observe waveforms such as those shown in Figure 2–2(a). These waveforms are typical outcomes if, for example, the random variable Θ is uniformly distributed on the interval $[0, 2\pi]$. On the other hand, if Θ takes on the values 0 and $\pi/2$ only, the two waveforms illustrated in Figure 2–2(b) are the *only* ones possible.

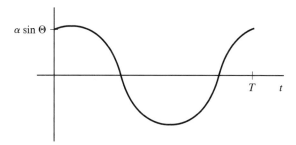

FIGURE 2–1 A sample function for $X(t) = \alpha \sin(2\pi f_0 t + \Theta)$.

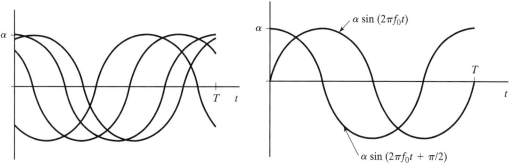

(a) Typical sample functions if Θ is uniform on $[0, 2\pi]$ (b) Possible sample functions if Θ takes values 0 and $\pi/2$ only

FIGURE 2–2 Typical sample functions for $X(t) = \alpha \sin(2\pi f_0 t + \Theta)$.

Notice that for each value of t, $X(t)$ is a function of the random variable Θ, and hence $X(t)$ is itself a random variable. We are interested in statistical characterizations of the random variable $X(t)$, including the distribution function for $X(t)$, and these characterizations depend on the distribution function for Θ. ∎

Example 2–2 Thermal Noise

Suppose we conduct an experiment in which a resistor of R ohms is placed in a controlled environment with the temperature held constant at T kelvins (K). As shown in Figure 2–3(a), the terminals of the resistor are connected to the input of an ideal band-pass filter (BPF) with center frequency f_0 Hz and bandwidth B Hz. (See Figure 2–3(b).)

The output $Y(t)$ of the band-pass filter obtained in this experiment will be a random-looking waveform that does not appear to exhibit much statistical regularity, as illustrated in Figure 2–4. If, however, we observe the average power $W(t)$, we will see that it always settles down to a value close to $w_0 = 4kTRB$, also illustrated in the figure, where $k \approx 1.38 \times 10^{-23}$ joule/K is Boltzmann's constant. With R in ohms and B in Hz, $w_0 = 4kTRB$ is in volts squared. The reason $W(t)$ is referred to as "power" is that it represents the power dissipated if $Y(t)$ is the voltage across a *one-ohm* resistor. The resulting power is then referred to as the *power on a one-ohm basis*. A more precise statement is that $\sqrt{W(t)}$ is the rms value, averaged from time 0 to time t, of the voltage $Y(t)$ at the filter output.

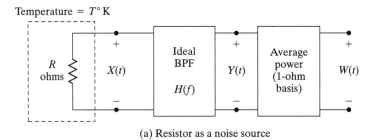

(a) Resistor as a noise source

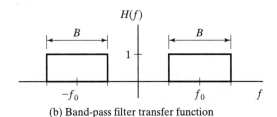

(b) Band-pass filter transfer function

FIGURE 2–3   Generation of thermal noise.

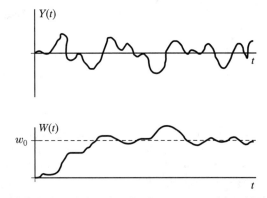

FIGURE 2–4   Sample functions for the processes $Y(t)$ and $W(t)$.

The phenomenon just described was observed experimentally by Johnson and derived analytically by Nyquist in 1928. The model is valid for frequencies f_0 up to at least 10^{11} Hz. For an idea of the magnitude of the numbers involved, consider the following typical values of the parameters: $T = 293$ K (approximately room temperature), $B = 10$ kHz, and $R = 10{,}000$ ohms. The result is that $w_0 \approx 1.6 \times 10^{-12}$ volts squared. This corresponds to an rms voltage of about 1.3 μV.

The experiment can be repeated several times, in which case a variety of different random-looking waveforms will be obtained at the output of the filter. The set of all such waveforms that could result from this experiment is the set of sample functions of the process $Y(t)$. In spite of the differences between different sample functions, the average power $W(t)$ will stay close to $w_0 = 4kTRB$ for sufficiently large t.

The foregoing discussion illustrates just one example in which the process of interest may appear not to exhibit any statistical regularity, but upon closer examination it will be seen to have well-defined second-order properties. In this example, the second-order property considered is the time-averaged power.   ■

Example 2–3 A Counting Process

Suppose that at time $t = 0$ we begin counting the number of phone calls coming into a switchboard. For $t > 0$, let $X(t)$ be the total number of phone calls received at or before time t. The counting process is initialized by setting $X(0) = 0$. If this experiment is repeated several times (e.g., once each day for several days) and the results are recorded in the form of graphs of $X(t)$ vs. t, the outcomes will be several different increasing step functions. A typical outcome is shown in Figure 2–5.

If we postulate that no more than one call can be received at a time, all of the steps (jump discontinuities) in the recorded graphs will be of size one. The randomness is in the location of the steps, which are the arrival times of the phone calls. For a fixed time interval $[0, T]$, both the total number of steps and the location of the steps are random, and both the number and location of the steps can be determined from a knowledge of $X(t)$ for all t in the range $0 \leq t \leq T$.   ■

Example 2–4 A Discrete-Time Filtering Problem

Suppose that $X_0, X_1, X_2, \ldots$ is a sequence of independent random variables whose distribution is defined by $P(X_n = 0) = P(X_n = 1) = \frac{1}{2}$ for each n. Define a discrete-time random process $X(t)$, $t = 0, 1, 2, \ldots$, by setting $X(t) = X_t$. Let $X(t)$ be the input to a discrete-time filter for which the output at time t is the sum of the input at time t and the input at time $t - 1$. Hence, the output is $Y(t) = X(t) + X(t - 1)$. The output process is then a discrete-time random process $Y(t)$ consisting of a sequence $Y_1, Y_2, Y_3, \ldots$ of random variables for which

$$P(Y_k = 0) = P(Y_k = 2) = \frac{P(Y_k = 1)}{2} = \frac{1}{4}$$

for each k. However, the sequence $Y_1, Y_2, Y_3, \ldots$ is *not* a sequence of independent random variables. For instance,

$$P(Y_1 = 2, Y_2 = 0) = 0 \neq P(Y_1 = 2)P(Y_2 = 0) = \frac{1}{16}.$$ ■

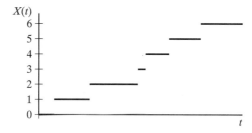

FIGURE 2–5   A typical sample function for the counting process.

2.2 DEFINITIONS AND BASIC CONCEPTS

Thus far, we have been using the term "random process" in an intuitive way, without a formal mathematical description. To proceed further, it is necessary to give a precise definition. A *random process* on a probability space $(\Omega, \mathcal{F}, P)$ is defined as an indexed collection $\{X_t : t \in \mathbb{T}\}$ of random variables on $(\Omega, \mathcal{F}, P)$. Random processes are also often called *stochastic processes* in the literature.

The parameter t typically denotes time, and the index sets of interest are usually intervals of the real line $\mathbb{R}$ (e.g., $\mathbb{T} = [0, \infty)$, $\mathbb{T} = (a, b)$, or $\mathbb{T} = (-\infty, \infty)$), subsets of set $\mathbb{Z}$ of all integers (e.g., $\mathbb{T} = \mathbb{Z}$, $\mathbb{T} = \{0, 1, 2, \dots\}$, or $\mathbb{T} = \{1, 2, \dots, n\}$), or sets that can be indexed by a subset of $\mathbb{Z}$ (e.g., $\mathbb{T} = \{1/n, 2/n, 3/n, \dots\}$, for some positive integer n). For convenience, in what follows, let $\mathbb{N}$ denote a set of consecutive integers. For example, $\mathbb{N}$ could be $\mathbb{Z}$, $\{0, 1, 2, \dots\}$, $\{1, 2, \dots, n\}$ for some integer n, or $\{-9, \dots, -1, 0, 1, \dots, 9\}$. These sets are all countable (i.e., they can be indexed by a subset of the integers), and they are ordered in the sense defined in Section 1.2.1. We refer to ordered countable sets as *discrete sets* in this book. An example of a set that is *not* countable is the set of all real numbers between 0 and 1.

Various types of random processes are encountered in engineering applications, and it is helpful to divide them into four categories, determined by the type of index set $\mathbb{T}$ and by the set of values that can be taken on by the random variables that make up the random processes. If $\mathbb{T}$ is an interval of the real line, $\{X_t : t \in \mathbb{T}\}$ is a *continuous-time* random process. If $\mathbb{T}$ is of the form $\{t_k : k \in \mathbb{N}\}$ and $t_k < t_i$ for all k and i in the set $\mathbb{N}$ such that $k < i$, then $\mathbb{T}$ is a *discrete-time* random process; that is, the index set for a discrete-time random process is a discrete set. In general, the set S of values taken on by a random process is called the *state space* of the process. Unless stated otherwise, the random processes in this book are assumed to be real valued; that is, $S \subset \mathbb{R}$. A real-valued random process is simply an indexed collection of real-valued random variables. If the state space of a random process is a discrete set, we say that the *state space is discrete* and the process is a *discrete-amplitude* random process. If $S \subset \mathbb{R}$, but S is not discrete, such as when S is an interval of $\mathbb{R}$, the process is said to be a *continuous-amplitude* random process.

Notice that the processes in Examples 2–1 and 2–2 are continuous-amplitude, continuous-time processes. The process in Example 2–3 is a discrete-amplitude, continuous-time process, and the process in Example 2–4 is a discrete-amplitude, discrete-time process. For an example of the only remaining type—a continuous-amplitude, discrete-time process—consider a random process formed from a sequence of independent Gaussian random variables.

The statement "$\{X_t : t \in \mathbb{T}\}$ is a collection of random variables on $(\Omega, \mathcal{F}, P)$" means that for each $t \in \mathbb{T}$, $\{X_t \leq u\} \in \mathcal{F}$. This latter condition is just shorthand notation for the requirement that $\{\omega \in \Omega : X_t(\omega) \leq u\}$ be an event for each real number u. (See Figure 2–6.)

Because X_t is a random variable for each t,

$$\{X_t \leq u\} = \{\omega \in \Omega : X_t(\omega) \leq u\} \in \mathcal{F}$$

for each choice of t and u. It follows that

$$\bigcap_{k=1}^{n} \{X_{t_k} \leq u_k\} \in \mathcal{F},$$

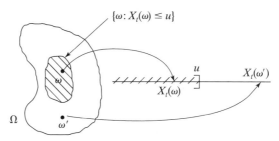

FIGURE 2–6   The event $\{\omega \in \Omega : X_t(\omega) \le u\}$.

so that probabilities of the form

$$P(X_{t_1} \le u_1, X_{t_2} \le u_2, \ldots, X_{t_n} \le u_n) = P\left[\bigcap_{k=1}^{n} \{X_{t_k} \le u_k\}\right]$$

are defined for each positive integer n; each set of points $t_1, t_2, \ldots, t_n$; and each choice of $u_1, u_2, \ldots, u_n$. Such probabilities are fundamental building blocks for the analysis of random processes in communication systems and other applications. We consider them further in Section 2.2.1.

From an engineering point of view, it is sometimes helpful to think of a continuous-time random process as a collection of waveforms and a discrete-time random process as a collection of sequences, as discussed in Section 2.1. Mathematically, this point of view is derived from the consideration of $X_t(\omega)$ as a function of t for fixed ω. We then refer to $X_t(\omega)$ as a *sample function* of the process. The notion of repeating an experiment several times, which is mentioned in the examples of Section 2.1, can be viewed as drawing several points at random from Ω according to the probability assignment P. If the outcomes of these drawings are $\omega_1, \omega_2, \ldots, \omega_k$, then the experimentally observed waveforms are $X_t(\omega_1), X_t(\omega_2), \ldots, X_t(\omega_k)$, each considered as a function of t. Hence, we can either think of drawing waveforms from a large collection of waveforms or drawing points from Ω and letting $X_t(\omega)$ generate the waveforms. Thus, on the kth drawing, the point ω_k is selected from Ω, and the sample function $X_t(\omega_k)$ is generated. This point of view is illustrated in Figure 2–7 for $k = 3$.

Thus, we may interpret a random process as a collection of waveforms that are indexed by the parameter $\omega \in \Omega$. This amounts to a reversal of the roles of "variable" and "parameter" from the point of view adopted in the original mathematical definition. In the definition of a random process given at the beginning of this section, the "parameter" is t and the "variable" is ω. In order to view a random process as a collection of waveforms, the "variable" becomes t and the "parameter" is ω, in which case it is more natural to use the notation $X(t, \omega)$ than $X_t(\omega)$.

Most engineering textbooks use either $X(t), t \in \mathbb{T}$, or simply $X(t)$, to denote a random process; that is, the variable ω in the notation $X(t, \omega)$ is suppressed. We adopt that convention in most of what follows, except for situations in which it is important to display the functional dependence on ω for clarity. In such cases, the notation will be $X(t, \omega)$ or $X_t(\omega)$, whichever is more appropriate. In using the engineering notation $X(t)$, the reader must keep in mind that for each t, $X(t)$ is a *random variable* on the probability space $(\Omega, \mathcal{F}, P)$, and its value at the point $\omega \in \Omega$ is $X(t, \omega)$. For a fixed value

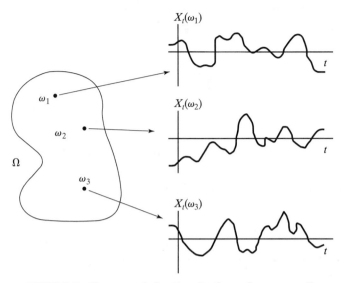

FIGURE 2–7   Three sample functions for the random process X_t.

of t, $X(t)$ is actually a function $X_t : \Omega \rightarrow \mathbb{R}$, a mapping from Ω into the real line, and this function satisfies $\{\omega \in \Omega : X(t, \omega) \leq u\} \in \mathcal{F}$ for each choice of t and u.

Example 2–5   A Ramp Signal with a Random Slope

Let the sample space Ω for the random process consist of all real numbers (i.e., $\Omega = \mathbb{R}$, the real line), and let the index set $\mathbb{T}$ be the set of nonnegative real numbers (i.e., $\mathbb{T} = [0, \infty)$). Suppose a random process is defined by $X_t(\omega) = X(t, \omega) = \omega t$ for each $\omega \in \mathbb{R}$ and each $t \geq 0$. For a fixed value of ω, $X_t(\omega)$ is a straight line of slope ω that starts at the origin. Some sample functions for this random process are shown in Figure 2–8.

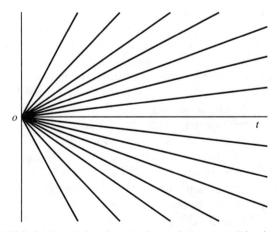

FIGURE 2–8   Sample functions for the random process $X(t, \omega) = \omega t$.

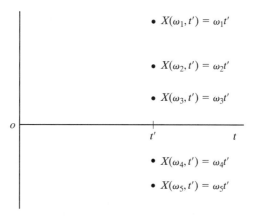

FIGURE 2–9   Sample values for the random processes $X(t, \omega) = \omega t$ at time t'.

As is true of any random process, the random process $X_t(\omega) = \omega t$ has the property that, for any fixed value of t, X_t is a random variable. We can think of fixing $t = t'$, drawing a point at random from Ω, and examining the value of $X_{t'}(\omega) = \omega t'$. This procedure is illustrated in Figure 2–9. Notice that what we observe will depend on the value of t' that was chosen. For example, "most" sample functions give a value of $\omega t'$ that is small in magnitude if t' is small, but "most" sample functions give a value of $\omega t'$ that is large in magnitude if the value of t' is large. A precise interpretation of "most" depends on the probability measure P on the sample space; probability distributions associated with random processes are discussed in the next subsection. ■

The two exercises that follow are optional. The concepts illustrated by these exercises are useful, but they are not required for the material presented elsewhere in the text.

Exercise 2–1. Suppose that $\Omega = (0, \infty)$ and we define $X(t, \omega)$ by

$$X(t, \omega) = \omega t$$

for all $t > 0$ and all $\omega \in \Omega$. This exercise is similar to Example 2–5, except that only positive slopes are allowed and we now have the restriction that $t > 0$. Suppose that $\mathcal{F}$ contains all intervals $(a, b]$ for which $0 \leq a < b$, and suppose the probability measure P is such that $P((a, b]) = e^{-a} - e^{-b}$. Show that $X(t, \omega)$ is a random process, and find $P(X(t) \leq u)$ for each $t > 0$ and each value of u.

Solution. To show that $X(t, \omega)$ is a random process, we let t be arbitrary, but fixed, and show that $\Omega_u \in \mathcal{F}$ for each u, where $\Omega_u = \{\omega \in \Omega : X(t, \omega) \leq u\}$. We then replace $X(t, \omega)$ by ωt in the expression for the set Ω_u and notice that, for $u > 0$,

$$\Omega_u = \{\omega : 0 < \omega t \leq u\} = \{\omega : 0 < \omega \leq t^{-1}u\} = (0, t^{-1}u],$$

and for $u \leq 0$,

$$\Omega_u = \{\omega : 0 < \omega t \leq u\} = \emptyset.$$

Now, $P(X(t) \leq u) = P(\Omega_u) = 1 - \exp(-t^{-1}u)$ for $u > 0$, and $P(X(t) \leq u) = P(\emptyset) = 0$ for $u \leq 0$. ■

Exercise 2–2. Suppose the amplitude of the oscillator output in Example 2–1 is $\alpha = 1$ and the random variable Θ is defined by $\Theta(\omega) = \omega$ for each ω in $\Omega = [0, 2\pi]$. Show that $X(t, \omega) = \sin[2\pi f_0 t + \Theta(\omega)]$ is a random process if $\mathcal{F}$ is an event class that includes all intervals of $[0, 2\pi]$.

Solution. Since, for $0 \leq t \leq T$,

$$X(t, \omega) = \sin[2\pi f_0 t + \omega], \quad \text{for } 0 \leq \omega \leq 2\pi,$$

it follows that the set $\Omega_u = \{\omega \in \Omega : X(t, \omega) \leq u\}$ is empty if $u < -1$ and Ω_u is just the entire sample space Ω if $u \geq 1$. On the other hand, if $-1 \leq u < 1$, the set

$$\Omega_u = \{\omega \in \Omega : \sin(2\pi f_0 t + \omega) \leq u\}$$

is a union of not more than two intervals, as illustrated in Figure 2–10 for two specific values of t: $t = 0$ and $t = t_1 = (4 f_0)^{-1}$. If $\mathcal{F}$ contains all intervals of $\Omega = [0, 2\pi]$, then it must also contain all countable unions of such intervals. Hence, $\Omega_u \in \mathcal{F}$ for each u. ∎

In engineering, we are concerned with the application of probabilistic methods to the solution of practical problems involving random processes. Consequently, we usually need not concern ourselves with the detailed mathematical structure of the random process. In particular, we can assume that each random phenomenon encountered in practice is a valid random process for some choice of the probability space $(\Omega, \mathcal{F}, P)$. Hence, we study the detailed structure of $X(t, \omega)$ only if it is convenient to do so and only when it is helpful in solving the problem at hand. For instance, an examination of the structure of the random process at the level of Exercises 2–1 and 2–2 is not necessary for the solution of the vast majority of engineering problems. Instead, we are usually able to derive all of the information we require from the finite-dimensional distribution functions, and these are the subject of the next subsection.

2.2.1   Distribution and Density Functions for Random Processes

The *one-dimensional distribution function* for a random process $X(t), t \in T$, is denoted by $F_{X,1}$ and defined by

$$F_{X,1}(u; t) = P(X(t) \leq u) = P(\{\omega : X(t, \omega) \leq u\})$$

for each real number u and each $t \in \mathbb{T}$. This distribution function is also known as the univariate distribution function for the random process. For each positive integer n, the *n-dimensional distribution function* is denoted by $F_{X,n}$ and defined by

$$F_{X,n}(u_1, \ldots, u_n; t_1, \ldots, t_n) = P[X(t_1) \leq u_1, \ldots, X(t_n) \leq u_n],$$

which is equivalent to

$$F_{X,n}(u_1, \ldots, u_n; t_1, \ldots, t_n) = P\left[\bigcap_{k=1}^{n} \{X(t_k) \leq u_k\}\right]$$
$$= P\left[\bigcap_{k=1}^{n} \{\omega : X(t_k, \omega) \leq u_k\}\right]$$

(2.1)

for each $u_k \in \mathbb{R}$ and $t_k \in \mathbb{T}$ ($1 \leq k \leq n$). For our purposes, a random process is specified completely by giving its n-dimensional distribution function for each positive integer n.

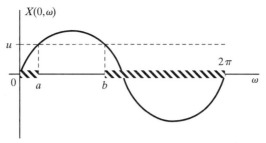

(a) $t = 0, 0 < u < 1, \Omega_u = [0, a] \cup [b, 2\pi], a = \sin^{-1}(u), b =$

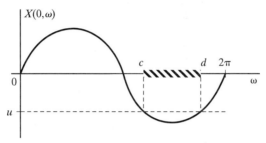

(b) $t = 0, -1 \leq u \leq 0, \Omega_u = [c, d], d = 2\pi + \sin^{-1}(u), c = 3\pi - d$

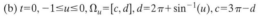

(c) $t = t_1 = (4f_0)^{-1}, 0 < u < 1, \Omega_u = [a, b], a = (\pi/2) - \sin^{-1}(u), b = 2\pi - a$

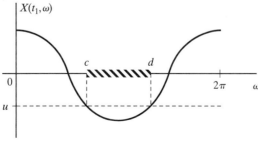

(d) $t = t_1 = (4f_0)^{-1}, -1 \leq u \leq 0, \Omega_u = [c, d], c = (\pi/2) - \sin^{-1}(u), d = 2\pi - c$

FIGURE 2–10 $\Omega_u = \{\omega \in [0, 2\pi] : \sin(2\pi f_0 t + \omega) \leq u\}.$

Just as with random variables, if each of the distribution functions $F_{X,n}$ has an associated density function $f_{X,n}$, then we say that each random vector $[X(t_1), \ldots, X(t_n)]$ is a *continuous random vector* with density $f_{X,n}$. This means that

$$F_{X,n}(x_1, \ldots, x_n; t_1, \ldots, t_n) = \int_{-\infty}^{x_n} \cdots \int_{-\infty}^{x_1} f_{X,n}(y_1, \ldots, y_n; t_1, \ldots, t_n)\, dy_1 \ldots dy_n \quad (2.2)$$

for each n and each choice of $\mathbf{x} = (x_1, \ldots, x_n)$ and $\mathbf{t} = (t_1, \ldots, t_n)$. It follows that we can let

$$f_{X,n}(y_1, \ldots, y_n; t_1, \ldots, t_n) = \partial^n F_{X,n}(x_1, \ldots, x_n; t_1, \ldots, t_n)/\partial x_1 \ldots \partial x_n|_{\mathbf{x}=\mathbf{y}}$$

for each $\mathbf{y} = (y_1, \ldots, y_n)$ for which the derivative exists.

Next, we consider a certain class of discrete-amplitude random processes: the class of random processes for which the state space $S = \{s_k : k \in \mathbb{N}\}$ is a discrete ordered set according to the definition in Section 1.2.1; that is, $s_i < s_j$ whenever $i < j$. Two important examples are $S = \mathbb{Z}$ and $S = \{0, 1, 2, \ldots\}$. For a random process in this class, we can always define a *discrete density function* (also known as a probability mass function) from the distribution function for the process. For example, the one-dimensional discrete density function for such a process $X(t)$ is given by

$$f_{X,1}(s_i; t) = F_{X,1}(s_i; t) - F_{X,1}(s_{i-1}; t),$$

and the two-dimensional discrete density is

$$f_{X,2}(s_i, s_j; t_1, t_2) = F_{X,2}(s_i, s_j; t_1, t_2) - F_{X,2}(s_{i-1}, s_j; t_1, t_2)$$
$$- F_{X,2}(s_i, s_{j-1}; t_1, t_2) + F_{X,2}(s_{i-1}, s_{j-1}; t_1, t_2).$$

On the other hand, for the n-dimensional discrete density function $f_{X,n}$, the corresponding distribution function is given by

$$F_{X,n}(s_{i_1}, \ldots, s_{i_n}; t_1, \ldots, t_n) = \sum_{j_1 \leq i_1} \cdots \sum_{j_n \leq i_n} f_{X,n}(s_{j_1}, \ldots, s_{j_n}; t_1, \ldots, t_n). \quad (2.3)$$

Of course, a random process cannot have both a density function [as in (2.2)] and a discrete density function [as in (2.3)]. Thus, in referring to the *density of a random process*, we mean the form of density that is appropriate for the particular random process in question. Obviously, for a discrete-amplitude process, only a discrete density function is appropriate, and the only density that can be considered for a continuous-amplitude process is a density of the form given in (2.2).

The solution to the following exercise illustrates how to determine the distribution and density functions for a discrete-amplitude random process.

Exercise 2–3. Find the one- and two-dimensional distribution and density functions for the discrete-amplitude, discrete-time process $Y(t)$ of Example 2–4.

Solution. The one-dimensional functions follow easily from the observation that, for $t \in \{1, 2, 3, \ldots\}$,

$$P[Y(t) = 0] = P[Y(t) = 2] = \frac{P[Y(t) = 1]}{2} = \frac{1}{4},$$

as noted in Example 2–4. Clearly, the discrete density function is given by

$$f_{Y,1}(y;t) = \begin{cases} \frac{1}{4}, & y = 0 \text{ or } y = 2, \\ \frac{1}{2}, & y = 1, \\ 0, & \text{otherwise}, \end{cases}$$

for $t \in \{1, 2, 3, \ldots\}$; thus, the distribution function is

$$F_{Y,1}(y;t) = \begin{cases} 0, & y < 0, \\ \frac{1}{4}, & 0 \le y < 1, \\ \frac{3}{4}, & 1 \le y < 2, \\ 1, & y \ge 2. \end{cases}$$

For two dimensions, we give only the density function, because the two-dimensional distribution function is somewhat cumbersome to describe for this example. The two-dimensional distribution function can, of course, be obtained from the two-dimensional density function via (2.3).

First we assume that t_1 and t_2 are positive integers for which $|t_1 - t_2| \ge 2$. This condition guarantees that

$$Y(t_1) = X(t_1) + X(t_1 - 1)$$

and

$$Y(t_2) = X(t_2) + X(t_2 - 1)$$

are independent. The condition $|t_1 - t_2| \ge 2$ implies that the values of $t_1, t_1 - 1, t_2$, and $t_2 - 1$ are all different, so the independence of $Y(t_1)$ and $Y(t_2)$ follows from the fact that the random process $X(t)$ of Example 2–4 has the property that $X(t)$ and $X(\tau)$ are independent if $t \ne \tau$. Thus, for $|t_1 - t_2| \ge 2$,

$$f_{Y,2}(y_1, y_2; t_1, t_2) = P[Y(t_1) = y_1, Y(t_2) = y_2] = P[Y(t_1) = y_1]P[Y(t_2) = y_2]$$
$$= f_{Y,1}(y_1; t_1) f_{Y,1}(y_2; t_2).$$

For $|t_1 - t_2| \le 1$, $Y(t_1)$ and $Y(t_2)$ are *dependent*. First notice that if $t_1 = t_2$, then

$$f_{Y,2}(y_1, y_2; t_1, t_2) = P[Y(t_1) = y_1, Y(t_1) = y_2] = \begin{cases} 0, & y_1 \ne y_2, \\ f_{Y,1}(y_1; t_1), & y_1 = y_2. \end{cases}$$

If $|t_1 - t_2| = 1$, then either $t_1 - 1 = t_2$ or $t_1 = t_2 - 1$ (but not both, of course), so we can write

$$Y(t_1) = Z_1 + Z_2$$

and

$$Y(t_2) = Z_2 + Z_3,$$

where Z_1, Z_2, and Z_3 are independent and satisfy $P(Z_i = 0) = P(Z_i = 1) = \frac{1}{2}$. If $t_1 - 1 = t_2$, then $Z_1 = X(t_1), Z_2 = X(t_2) = X(t_1 - 1)$, and $Z_3 = X(t_2 - 1)$. If $t_1 = t_2 - 1$, then $Z_2 = X(t_1) = X(t_2 - 1), Z_1 = X(t_1 - 1)$, and $Z_3 = X(t_2)$. In either case, we see that $P[Y(t_1) = y_1, Y(t_2) = y_2]$ is equal to $\frac{1}{8}$ for all pairs (y_1, y_2) in the set

$$S_1 = \{(0,0), (2,2), (0,1), (1,0), (2,1), (1,2)\},$$

is equal to $\frac{1}{4}$ for $y_1 = y_2 = 1$, and is equal to zero otherwise. Thus, for $|t_1 - t_2| = 1$,

$$f_{Y,2}(y_1, y_2; t_1, t_2) = \begin{cases} \frac{1}{8}, & (y_1, y_2) \in S_1, \\ \frac{1}{4}, & y_1 = y_2 = 1, \\ 0, & \text{otherwise.} \end{cases}$$

In particular, note that since it is impossible that both $Y(t) = 0$ and $Y(t + 1) = 2$ or both $Y(t) = 2$ and $Y(t + 1) = 0$, it must be that, for $|t_1 - t_2| = 1$,

$$f_{Y,2}(0, 2; t_1, t_2) = 0 = f_{Y,2}(2, 0; t_1, t_2),$$

and hence,

$$f_{Y,2}(y_1, y_2; t_1, t_2) \neq f_{Y,1}(y_1; t_1) f_{Y,1}(y_2; t_2) = \frac{1}{16}$$

whenever $y_1 = 0$ and $y_2 = 2$ (or $y_1 = 2$ and $y_2 = 0$). Clearly, $Y(t_1)$ and $Y(t_2)$ are statistically dependent, in contrast to the previous results for $|t_1 - t_2| \geq 2$.   ∎

2.2.2 *Gaussian Random Processes*

Next, we consider a very important class of continuous-amplitude random processes known as Gaussian random processes. A random process $X(t)$, $t \in \mathbb{T}$, is a *Gaussian random process* if, for any positive integer n, any choice of real coefficients a_k $(1 \leq k \leq n)$ and any choice of sampling times $t_k \in \mathbb{T}$ $(1 \leq k \leq n)$, the random variable $a_1 X(t_1) + a_2 X(t_2) + \cdots + a_n X(t_n)$ is a Gaussian random variable. As a consequence, for any $t_k \in \mathbb{T}$ $(1 \leq k \leq n)$, the vector

$$\mathbf{X}(t) = [X(t_1), \ldots, X(t_n)]$$

has the n-dimensional Gaussian density. In particular, the one-dimensional density function for a Gaussian random process is the Gaussian density function. The solution to the next exercise illustrates some methods for working with Gaussian random processes.

Exercise 2–4. Consider the continuous-amplitude, continuous-time random process $X(t)$, $t \in \mathbb{R}$, defined by

$$X(t) = Y_1 + tY_2,$$

where Y_1 and Y_2 are independent Gaussian random variables, each having zero mean and variance σ^2. Find the one- and two-dimensional density functions for this random process.

Solution. For each fixed t, $X(t)$ is a linear combination of the independent Gaussian random variables Y_1 and Y_2, so $X(t)$ is a Gaussian random variable. The mean of the random variable $X(t)$ is zero and the variance of $X(t)$ is

$$\text{Var}\{Y_1\} + \text{Var}\{tY_2\} = \sigma^2 + t^2\sigma^2 = \sigma^2(1 + t^2).$$

Hence,

$$f_{X,1}(x; t) = [2\pi\sigma^2(1 + t^2)]^{-1/2} \exp\{-x^2/[2\sigma^2(1 + t^2)]\}$$

is the one-dimensional density function for the process $X(t)$.

For fixed t_1 and t_2, $X(t_1)$ and $X(t_2)$ are jointly Gaussian with zero means and with variances given by

$$\text{Var}\{X(t_k)\} = \sigma^2(1 + t_k^2) \tag{2.4}$$

for $k = 1, 2$. Because the means are zero, the covariance is

$$\text{Cov}\{X(t_1), X(t_2)\} = E\{X(t_1)X(t_2)\}$$
$$= E\{(Y_1 + t_1 Y_2)(Y_1 + t_2 Y_2)\}$$
$$= \sigma^2 + t_1 t_2 \sigma^2 = \sigma^2(1 + t_1 t_2).$$

The correlation coefficient for $X(t_1)$ and $X(t_2)$ is defined by

$$\rho(t_1, t_2) = \frac{\text{Cov}\{X(t_1), X(t_2)\}}{\sqrt{\text{Var}\{X(t_1)\}\,\text{Var}\{X(t_2)\}}}.$$

For the random process given in this example, the correlation coefficient reduces to

$$\rho(t_1, t_2) = \frac{1 + t_1 t_2}{[(1 + t_1^2)(1 + t_2^2)]^{1/2}}. \tag{2.5}$$

Therefore, the two-dimensional density function for the process $X(t)$ is given by

$$f_{X,2}(x_1, x_2; t_1, t_2) =$$
$$\frac{\exp\{-[(1 + t_2^2)x_1^2 - 2(1 + t_1 t_2)x_1 x_2 + (1 + t_1^2)x_2^2]/[2\sigma^2(t_1 - t_2)^2]\}}{2\pi\sigma^2|t_1 - t_2|},$$

provided that $t_1 \neq t_2$. If $t_1 = t_2$, then the joint distribution of $X(t_1)$ and $X(t_2)$ is degenerate, since $\rho(t_1, t_1) = 1$. ∎

For the random process $X(t)$ of Exercise 2–4, we can show directly that the random variable

$$Z = \sum_{k=1}^{n} a_k X(t_k) = a_1 X(t_1) + \cdots + a_n X(t_n)$$

is a Gaussian random variable (without appealing to the definition of a Gaussian random process). First observe that Z can be expressed as $c_1 Y_1 + c_2 Y_2$, where

$$c_1 = \sum_{k=1}^{n} a_k \qquad \text{and} \qquad c_2 = \sum_{k=1}^{n} a_k t_k.$$

That is, Z is a linear combination of two jointly Gaussian random variables Y_1 and Y_2, and therefore, Z is Gaussian.

Some additional discussion of Gaussian random processes is given in Section 2.7. Many of the properties of Gaussian random processes that are used in this book are derived from the fact that each of the n-dimensional distribution functions ($n = 1, 2, 3, \ldots$) is completely specified by the appropriate means and covariances. (The variance is a special case of the covariance; e.g., $\text{Var}\{X(t)\} = \text{Cov}\{X(t), X(t)\}$.) The reader should observe that the foregoing solution to Exercise 2–4 demonstrates this fact for the two-dimensional distribution of the Gaussian random process in question.

2.2.3 Conditional Densities for Random Processes

Once we have defined n-dimensional density functions, we can obtain conditional density functions in a straightforward manner, since conditional density functions are just ratios of joint density functions. We shall engage in a slight abuse of notation at this point by using the same notation $f_{X,n}$ for both the conditional and joint densities. The appearance of a vertical bar (e.g., as in $f_{X,n}(\cdot|\cdot)$) will, as usual, inform the reader that the function should be interpreted as a *conditional* density.

If, for each k $(1 \leq k \leq n)$, the function $f_{X,k}$ is the k-dimensional density function for a continuous-amplitude random process $X(t)$, then the conditional density function for $X(t_1), \ldots, X(t_m)$ given $X(t_{m+1}), \ldots, X(t_n)$ is

$$f_{X,n}(x_1, \ldots, x_m; t_1, \ldots, t_m | x_{m+1}, \ldots, x_n; t_{m+1}, \ldots, t_n) =$$
$$\frac{f_{X,n}(x_1, \ldots, x_n; t_1, \ldots, t_n)}{f_{X,n-m}(x_{m+1}, \ldots, x_n; t_{m+1}, \ldots, t_n)} \tag{2.6}$$

for $n \geq 2$ and $1 \leq m < n$.

Exercise 2–5. In Exercise 2–4, find the conditional density function for $X(t_1)$ given $X(t_2)$ if $t_1 \neq t_2$.

Solution. By simply substituting into the expression

$$f_{X,2}(x_1; t_1 | x_2; t_2) = \frac{f_{X,2}(x_1, x_2; t_1, t_2)}{f_{X,1}(x_2; t_2)}$$

for the densities $f_{X,2}$ and $f_{X,1}$ determined in Exercise 2–4, we find that

$$f_{X,2}(x_1; t_1 | x_2; t_2) = \frac{\exp\{-[x_1(1 + t_2^2) - (1 + t_1 t_2)x_2]^2/2\sigma^2(t_1 - t_2)^2(1 + t_2^2)\}}{[2\pi\sigma^2(t_1 - t_2)^2/(1 + t_2^2)]^{1/2}}. \tag{2.7}$$

Alternatively, we can simply use the fact that $X(t_1)$ and $X(t_2)$ are jointly Gaussian, so that $f_{X,n}(\cdot|\cdot)$ has the form of a Gaussian density. The conditional mean is

$$\alpha = E\{X(t_1)|X(t_2) = x_2\} = \left[\frac{\mathrm{Var}\{X(t_1)\}}{\mathrm{Var}\{X(t_2)\}}\right]^{1/2} \rho(t_1, t_2)x_2$$

(since $E\{X(t_1)\} = E\{X(t_2)\} = 0$) and the conditional variance is

$$\beta^2 = \mathrm{Var}\{X(t_1)|X(t_2)\} = \mathrm{Var}\{X(t_1)\}(1 - [\rho(t_1, t_2)]^2).$$

Substituting into these two expressions from (2.4) and (2.5), we have

$$\alpha = \left[\frac{1 + t_1 t_2}{1 + t_2^2}\right] x_2$$

and

$$\beta^2 = \frac{\sigma^2(t_1 - t_2)^2}{1 + t_2^2}.$$

Since the conditional density is then of the form

$$\frac{\exp\{-[x_1 - \alpha]^2/2\beta^2\}}{\sqrt{2\pi\beta^2}},$$

we arrive at (2.7) after some rearrangement of terms.   ∎

2.3 MEAN, AUTOCORRELATION, AND AUTOCOVARIANCE

It is not always practical (or even possible) to determine all of the n-dimensional distribution functions for a random process. Furthermore, it is usually unnecessary to have that much information about the process in order to obtain a satisfactory solution to an engineering problem. In most of the problems encountered in engineering, a great deal can be learned from the two-dimensional distribution functions. The most important "parameters" for the two-dimensional distributions are the mean, autocorrelation, and autocovariance functions.

2.3.1 Definitions and Basic Properties

We define the mean, autocorrelation, and autocovariance functions for random processes $X(t)$ that satisfy $E\{[X(t)]^2\} < \infty$ for each t. Such processes are called *second-order* processes. All processes in this section are assumed to be second-order, real-valued random processes, unless stated otherwise. If $X(t)$ represents a physical signal, such as voltage or current, in an electrical system, $E\{[X(t)]^2\}$ represents the expected value of the instantaneous power (on a one-ohm basis) in the process $X(t)$. When the random process $X(t)$ arises in such a manner—and this is the situation of primary interest to us in electrical engineering—it is reasonable to assume that $E\{[X(t)]^2\}$ is finite. In the next exercise, we establish that $E\{|X(t)|\}$ is also finite for a second-order process $X(t)$. In this exercise and for the remainder of the book, we omit the braces from the expressions involving expectations whenever the meaning is clear. Thus, $E\{[X(t)]^2\}$ is written more simply as $E[X(t)]^2$. Similarly, $E\{X(t)\}$ and $E\{|X(t)|\}$ are often denoted by $EX(t)$ and $E|X(t)|$, respectively. In addition, the index set $\mathbb{T}$ is frequently not displayed explicitly; in such cases, statements such as "for each t" should be interpreted as "for each t in $\mathbb{T}$."

Exercise 2–6. Show that a second-order random process $X(t)$ always satisfies $E|X(t)| < \infty$ for each t.

Solution. First observe that for any real number u, $|u| \leq 1 + u^2$. The easiest way to see this is to consider the two cases $|u| < 1$ and $|u| \geq 1$ separately. If $|u| < 1$, then $|u|$ is certainly no larger than $1 + u^2$, and if $|u| \geq 1$ then $|u| \leq u^2$. An application of this inequality gives

$$|X(t)| \leq 1 + [X(t)]^2. \tag{2.8}$$

Therefore, for each t,

$$E|X(t)| \leq 1 + E[X(t)]^2 < \infty.$$

We should point out that what is really meant by (2.8) is that

$$|X(t, \omega)| \leq 1 + [X(t, \omega)]^2$$

for each t and each ω in Ω. Following the standard convention in engineering, we have omitted the ω-dependence in (2.8), since the meaning is clear. ∎

The *mean function* (or simply the *mean*) of a second-order random process $X(t)$, $t \in \mathbb{T}$, is defined by

$$\mu_X(t) = E\{X(t)\}$$

for each $t \in \mathbb{T}$. The *autocorrelation function* is defined by

$$R_X(t_1, t_2) = E\{[X(t_1)\, X(t_2)]\}$$

for t_1 and t_2 in $\mathbb{T}$, and the *autocovariance function* is defined by

$$C_X(t_1, t_2) = E\{[X(t_1) - \mu_X(t_1)][X(t_2) - \mu_X(t_2)]\}.$$

As will be shown later, $R_X(t_1, t_2)$ and $C_X(t_1, t_2)$ are finite for a second-order process.

In a sense, the mean $\mu_X(t)$ is the average value of the random process at time t, where "average" refers to the probabilistic average or expectation. The autocorrelation $R_X(t_1, t_2)$ and autocovariance $C_X(t_1, t_2)$ are quantitative measures of a certain type of statistical "coupling" between $X(t_1)$ and $X(t_2)$. For instance, if $X(t_1)$ and $X(t_2)$ are *statistically independent*, there is no such "coupling," and, in fact, $C_X(t_1, t_2) = 0$, which is equivalent to saying that the two random variables $X(t_1)$ and $X(t_2)$ are uncorrelated.

The autocorrelation and autocovariance functions play a very important role in the analysis of random processes in linear systems. One of the important topics in this subject is the study of noise in linear systems, and one of the most fundamental parameters of noise is the power in the noise. This is needed, for instance, in order to determine the signal-to-noise ratios at the input and output of a linear system. A filter is often employed in an electronic system to decrease the noise power (i.e., increase the signal-to-noise ratio). In such applications, it is important to be able to obtain analytical expressions for the noise power at the output of the filter as a function of the filter characteristics. The results developed in Chapters 2–4 will enable the reader to obtain such expressions, which can then be used to perform engineering tradeoff studies in the design of the filter.

If the random process $X(t)$ represents noise, then, as we have discussed, the power in the noise is $E[X(t)]^2$. From the definition of the autocorrelation function, we see that this power is $R_X(t, t)$. Suppose $X(t)$ is the noise at the input to a linear system, and ignore the signal for the present discussion. In this case, the output $Y(t)$ is also a random process (i.e., noise in gives noise out). In order to determine the output noise power $E[Y(t)]^2$, it turns out that we need to know more than the input noise power. In fact, except for trivial linear systems, it is necessary to know the autocorrelation function $R_X(t_1, t_2)$ for the input random process for several values of its arguments t_1 and t_2, even if all we want to know about the output noise process is its power.

Generally, the mean, autocorrelation, and autocovariance functions are relatively easy to determine, compared with other parameters of a random process. This is because these functions can all be determined from the two-dimensional distribution function $F_{X,2}$. In fact, the mean can be determined even if all that we know is the one-dimensional distribution function. Notice that

$$C_X(t_1, t_2) = R_X(t_1, t_2) - \mu_X(t_1)\mu_X(t_2), \tag{2.9}$$

so that it suffices to consider the mean together with only *one* of the two functions R_X or C_X. Notice also that $C_X(t_1, t_2) = 0$ if and only if $R_X(t_1, t_2) = \mu_X(t_1)\mu_X(t_2)$.

Next, we establish that the mean, autocorrelation, and autocovariance functions are finite for a process with finite power. First we use the fact that

$$-|X(t)| \leq X(t) \leq |X(t)|,$$

which implies that both $+EX(t)$ and $-EX(t)$ are not greater than $E|X(t)|$, so that

$$|\mu_X(t)| = |EX(t)| \le E|X(t)|. \tag{2.10}$$

This fact and Exercise 2–6 establish that $\mu_X(t)$ is finite for any second-order process. Also, we will see in Exercise 2–7 that

$$|R_X(t_1, t_2)| \le \frac{E[X(t_1)]^2 + E[X(t_2)]^2}{2}, \tag{2.11}$$

so that $R_X(t_1, t_2)$ is also finite for any second-order process. Finally, observe that (2.9) implies that $C_X(t_1, t_2)$ is finite whenever the mean and autocorrelation functions are finite, so the autocovariance function is finite for any second-order process.

Exercise 2–7. Show that

$$|R_X(t_1, t_2)| \le \frac{R_X(t_1, t_1) + R_X(t_2, t_2)}{2}. \tag{2.12}$$

Notice that (2.11) and (2.12) are the same inequality written in different notation.

Solution. Since the expected value of the square of a random variable is nonnegative, the following inequalities are valid for all t_1 and t_2:

$$0 \le E[X(t_1) + X(t_2)]^2 = R_X(t_1, t_1) + 2R_X(t_1, t_2) + R_X(t_2, t_2); \tag{2.13}$$

$$0 \le E[X(t_1) - X(t_2)]^2 = R_X(t_1, t_1) - 2R_X(t_1, t_2) + R_X(t_2, t_2). \tag{2.14}$$

But these two inequalities imply that both $+R_X(t_1, t_2)$ and $-R_X(t_1, t_2)$, and hence $|R_X(t_1, t_2)|$, are not greater than the right-hand side of (2.12). ∎

At the beginning of this subsection, it was pointed out that if $X(t_1)$ and $X(t_2)$ are *independent*, the autocovariance function satisfies $C_X(t_1, t_2) = 0$. If, on the other hand, $X(t_1) = X(t_2)$, which is, intuitively, as dependent as two random variables can be, then

$$C_X(t_1, t_2) = E[X(t_1) - \mu_X(t_1)]^2 = E[X(t_2) - \mu_X(t_2)]^2$$
$$= \text{Var}\{X(t_1)\} = \text{Var}\{X(t_2)\}.$$

In order to explore this relationship further, fix t_1 and t_2, and let

$$\sigma_i = \sqrt{\text{Var}\{X(t_i)\}}$$

for $i = 1$ and $i = 2$. Now, define the *normalized autocovariance function* K_X by

$$K_X(t_1, t_2) = \frac{C_X(t_1, t_2)}{\sigma_1 \sigma_2}.$$

Later in this section it is shown that $|C_X(t_1, t_2)| \le \sigma_1 \sigma_2$, so that the normalized autocovariance function is always between -1 and $+1$. Notice that our example $X(t_1) = X(t_2)$ gives the *largest* possible positive normalized autocovariance, and if $X(t_1)$ and $X(t_2)$ are independent, we get $K_X(t_1, t_2) = 0$, the smallest possible magnitude for the normalized autocovariance. Notice also that the latter equation is equivalent to $C_X(t_1, t_2) = 0$, which, from (2.9), is in turn equivalent to $R_X(t_1, t_2) = \mu_X(t_1)\mu_X(t_2)$. As a side issue, note that the value -1, the smallest possible value for the normalized autocovariance, is achieved if $X(t_1) = -X(t_2)$. An example of a random process $X(t)$

with the property that $X(t_1) = X(t_2)$ for certain values of t_1 and t_2 is the sinusoid with a random phase angle, as described in Example 2–1: simply let $t_1 = 0$ and $t_2 = T$. Similarly, letting $t_1 = 0$ and $t_2 = T/2$ in this example gives $X(t_1) = -X(t_2)$. The full autocorrelation function for the sinusoidal signal with a random phase angle is derived in Exercise 2–8.

Often, we are dealing with random processes that have rather simple descriptions, such as the sinusoid with a random phase. In such cases, it is usually a straightforward exercise to determine the mean, autocorrelation, and autocovariance functions. In addition to Example 2–1, this is also the situation for Examples 2–4 and 2–5. The mean and autocorrelation functions for a sinusoidal signal with a random phase angle are found in the next exercise.

Exercise 2–8. Find the mean and autocorrelation function for the random process

$$X(t) = \alpha \sin(2\pi f_0 t + \Theta),$$

which was introduced in Example 2–1. The amplitude α is a deterministic constant, but the phase Θ is a random variable.

Solution. Let $\omega_0 = 2\pi f_0$, and observe that

$$X(t) = \alpha[\sin \Theta \cos \omega_0 t + \cos \Theta \sin \omega_0 t].$$

Thus, the mean of $X(t)$ is given by

$$\mu_X(t) = \alpha[E\{\sin \Theta\} \cos \omega_0 t + E\{\cos \Theta\} \sin \omega_0 t].$$

The autocorrelation function is

$$R_X(t_1, t_2) = \alpha^2 E\{\sin(\omega_0 t_1 + \Theta) \sin(\omega_0 t_2 + \Theta)\}$$
$$= \frac{\alpha^2}{2}[E\{\cos \omega_0(t_1 - t_2) - \cos[\omega_0(t_1 + t_2) + 2\Theta]\}].$$

It follows that

$$R_X(t_1, t_2) = \left(\frac{\alpha^2}{2}\right)[\cos[\omega_0(t_1 - t_2)] + E\{\sin 2\Theta\} \sin[\omega_0(t_1 + t_2)]$$
$$- E\{\cos 2\Theta\} \cos[\omega_0(t_1 + t_2)]].$$

This is the general result. Now suppose that Θ is uniform on $[0, 2\pi]$. In that case,

$$E\{\sin \Theta\} = E\{\cos \Theta\} = E\{\sin 2\Theta\} = E\{\cos 2\Theta\} = 0,$$

which is a much weaker condition on the distribution of Θ than is being uniformly distributed on $[0, 2\pi]$. It follows from the preceding condition that $\mu_X(t) = 0$ for each t and

$$R_X(t_1, t_2) = \left(\frac{\alpha^2}{2}\right) \cos[\omega_0(t_1 - t_2)]$$

for each t_1 and t_2. In particular, if $t_1 = 0$ and $t_2 = T$, it follows that

$$R_X(0, T) = \left(\frac{\alpha^2}{2}\right) \cos(\omega_0 T).$$

Using the fact that $T = f_0^{-1}$, we see that $\omega_0 T = 2\pi$, so $R_X(0, T) = \alpha^2/2$. Notice that $R_X(t, t) = \alpha^2/2$ for *any* time t, which can be deduced from the expression

$$R_X(t_1, t_2) = \left(\frac{\alpha^2}{2} \right) \cos[\omega_0(t_1 - t_2)]$$

by setting $t_1 = t_2 = t$. The fact that $R_X(t, t) = \alpha^2/2$ should not be surprising, because $R_X(t, t)$ is $E\{[X(t)]^2\}$, the expected value of the instantaneous power in the random process at time t. The average power in a sinusoidal signal of amplitude α is $\alpha^2/2$.

An interesting feature of this random process is that the power at time t need not be $\alpha^2/2$ for all choices of t if the distribution of the random phase angle Θ does not satisfy

$$E\{\sin \Theta\} = E\{\cos \Theta\} = E\{\sin 2\Theta\} = E\{\cos 2\Theta\} = 0.$$

For example, if $P(\Theta = 0) = 1$, then

$$E\{[X(0)]^2\} = E\{[X(T/2)]^2\} = E\{[X(T)]^2\} = 0,$$

which results from the expression for $X(t)$ if $\Theta = 0$ (or it can be seen from Figure 2–2). Notice that if Θ is uniformly distributed on $[0, 2\pi]$, then because the random process $X(t)$ has zero mean, $R_X(t_1, t_2) = C_X(t_1, t_2)$, and the normalized autocovariance is given by

$$K_X(t_1, t_2) = \frac{2R_X(t_1, t_2)}{\alpha^2} = \cos \omega_0(t_1 - t_2).$$

As is true in general, the maximum magnitude of the normalized autocovariance is unity. For this particular process, the maximum is achieved for several choices of t_1 and t_2, such as if $t_1 = t_2$ for any t_2 in the range $0 \le t_2 \le T$ or if t_1 and t_2 satisfy $|t_2 - t_1| = T/2$. Another choice is $t_1 = 0$ and $t_2 = T$ (i.e., $K_X(0, T) = +1$). ∎

The next exercise gives a simple illustration of a very important general procedure. The setting is as follows: The random process $X(t)$ is the input to a linear system and the random process $Y(t)$ is the output. Given some information about the input process, we wish to determine certain information about the output process. Examples of the type of information we are given or wish to determine are the mean, second moment, variance, autocorrelation function, and autocovariance function.

Exercise 2–9. Find the mean, autocorrelation, and autocovariance functions for the discrete-amplitude, discrete-time process $Y(t)$ defined in Example 2–4. Recall that here $Y(t) = X(t) + X(t - 1)$; $X(t) = X_t$ for $t = 0, 1, 2, \ldots$; and $X_0, X_1, X_2, \ldots$ is a sequence of independent random variables for which $P(X_t = 0) = P(X_t = 1) = \frac{1}{2}$ for each nonnegative integer t.

Solution. First notice that $E\{X_t\} = \frac{1}{2}$ and $E\{[X_t]^2\} = \frac{1}{2}$ for each nonnegative integer t. Since $Y(t) = X(t) + X(t - 1)$ for $t = 1, 2, 3, \ldots$, the mean function for $Y(t)$ is

$$\mu_Y(t) = E\{Y(t)\} = E\{X(t) + X(t - 1)\} = \mu_X(t) + \mu_X(t - 1).$$

But $\mu_X(t) = E\{X_t\} = \frac{1}{2}$ for all t, so $\mu_Y(t) = 1$ for all t. The autocorrelation function for $Y(t)$ is given by

$$\begin{aligned} R_Y(t_1, t_2) &= E\{[X(t_1) + X(t_1 - 1)][X(t_2) + X(t_2 - 1)]\} \\ &= R_X(t_1, t_2) + R_X(t_1, t_2 - 1) + R_X(t_1 - 1, t_2) + R_X(t_1 - 1, t_2 - 1). \end{aligned}$$

Because $X(t)$ and $X(s)$ are independent whenever $t \neq s$, $R_X(s,t) = \mu_X(t)\mu_X(s) = \frac{1}{4}$ for $t \neq s$. Furthermore, $R_X(s,t) = E[X(t)]^2 = \frac{1}{2}$ for $s = t$. Consequently, the expression for $R_Y(t_1, t_2)$ reduces to

$$R_Y(t_1, t_2) = \begin{cases} \frac{3}{2}, & |t_1 - t_2| = 0, \\ \frac{5}{4}, & |t_1 - t_2| = 1, \\ 1, & |t_1 - t_2| \geq 2. \end{cases}$$

Notice that $R_Y(t_1, t_2)$ can also be evaluated directly from the two-dimensional density function obtained in Exercise 2–3. We simply use the fact that

$$R_Y(t_1, t_2) = E[Y(t_1)Y(t_2)]$$

$$= \sum_{i=0}^{2} \sum_{j=0}^{2} ij f_{Y,2}(i, j; t_1, t_2).$$

By applying (2.9), we find that the autocovariance function for $Y(t)$ is given by

$$C_Y(t_1, t_2) = \begin{cases} \frac{1}{2}, & |t_1 - t_2| = 0, \\ \frac{1}{4}, & |t_1 - t_2| = 1, \\ 0, & |t_1 - t_2| \geq 2. \end{cases}$$

Alternatively, we can evaluate

$$C_Y(t_1, t_2) = \sum_{i=0}^{2} \sum_{j=0}^{2} (i - 1)(j - 1) f_{Y,2}(i, j; t_1, t_2).$$

Notice that the autocorrelation and autocovariance functions for the process $Y(t)$ do not depend on t_1 and t_2 individually, but only on their difference. Such processes are very important in the study of random processes in linear systems and are discussed further in Section 2.4 and in Chapters 3 and 4. Notice also that the autocorrelation and autocovariance functions are symmetric functions of t_1 and t_2 in the sense that t_1 and t_2 can be interchanged without changing the value of the function. We will see that this latter property is true for any real-valued random process. ∎

Recall that, for a given second-order random process $X(t), t \in \mathbb{T}$, the autocorrelation function for the process is a function of two variables:

$$R_X(t_1, t_2) = E\{X(t_1)X(t_2)\}.$$

However, not all functions of two variables can be an autocorrelation function for some random process. The fact that an autocorrelation function is a very special function has important implications in the analysis of second-order random processes. First we present some of the important properties of an autocorrelation function for a given second-order random process, and then we discuss the necessary and sufficient conditions for a given function of two variables to be a valid autocorrelation function for some second-order random process.

Given any second-order random process $X(t), t \in \mathbb{T}$, the autocorrelation function for $X(t)$ satisfies

$$R_X(t, t) \geq 0, \tag{2.15}$$

$$R_X(t_1, t_2) = R_X(t_2, t_1), \tag{2.16}$$

and

$$|R_X(t_1, t_2)| \le \sqrt{R_X(t_1, t_1)R_X(t_2, t_2)}, \qquad (2.17)$$

for each choice of t, t_1, and t_2. If (2.16) is satisfied for each t_1 and t_2, the function $R_X(\cdot, \cdot)$ is said to be *symmetric*.

In addition to satisfying (2.15)–(2.17), $R_X(\cdot, \cdot)$ is *nonnegative definite*; that is, for *any* positive integer n, *any* points $t_1, \ldots, t_n$ in $\mathbb{T}$, and *any* complex numbers $\alpha_1, \ldots, \alpha_n$,

$$\sum_{i=1}^{n} \sum_{k=1}^{n} \alpha_i \alpha_k^* R_X(t_i, t_k) \ge 0, \qquad (2.18)$$

where α_k^* denotes the complex conjugate of α_k. By (2.18) we mean, of course, that the indicated sum is *real* and *nonnegative*. An alternative term for nonnegative definite is *positive semidefinite*, and an equivalent definition to (2.18) is that, for any n and any $t_1, \ldots, t_n$ in $\mathbb{T}$, the $n \times n$ matrix that has the element $R_X(t_i, t_k)$ in the ith row and kth column is a nonnegative definite matrix.

Actually, it turns out that nonnegative definiteness implies (2.15)–(2.17), as is shown in Section 2.3.2. Moreover, it can be shown that, given a function $R(\cdot, \cdot)$ that is nonnegative definite, there always exists a second-order random process that has $R(\cdot, \cdot)$ as its autocorrelation function. Thus, nonnegative definiteness is a necessary and sufficient condition for a function $R(\cdot, \cdot)$ to be an autocorrelation function for some second-order random process.

Autocovariance functions must also satisfy all of the properties of autocorrelation functions. The reason is that, given a second-order process $X(t)$, the autocovariance function for $X(t)$ is just the autocorrelation function for the process $Y(t) = X(t) - \mu_X(t)$. Hence, any autocovariance function is an autocorrelation function. Furthermore, it can be shown that any symmetric, nonnegative definite function $C(\cdot, \cdot)$ is an autocovariance function for some second-order process. Consequently, the collection of all autocovariance functions is identical to the collection of all autocorrelation functions.

Exercise 2–10. Show that the following functions are not valid autocorrelation functions:

(a) $R(t_1, t_2) = |t_1 - t_2| \exp\{-3|t_1 - t_2|\}$, $t_1 \in \mathbb{R}$, $t_2 \in \mathbb{R}$;

(b) $R(t_1, t_2) = \begin{cases} 1, & |t_1 - t_2| \le T, \\ 2 - (|t_1 - t_2|/T), & T < |t_1 - t_2| \le 2T, \\ 0, & |t_1 - t_2| > 2T. \end{cases}$

Solution. The function in part (a) cannot be an autocorrelation function, because it does not satisfy (2.17). Notice that $R(t_1, t_1) = R(t_2, t_2) = 0$, but $R(t_1, t_2) > 0$, except for the special case $t_1 = t_2$.

The function in part (b) is not nonnegative definite. [Notice, however, that it does satisfy (2.15)–(2.17).] For instance, suppose $n = 3, t_1 = 0, t_2 = T, t_3 = 2T, \alpha_1 = \alpha_3 = +1$, and $\alpha_2 = -1$. For these values of the parameters, (2.18) is not satisfied. Alternatively, for the same choice of t_1, t_2, and t_3, we can simply observe that the matrix

$$\begin{bmatrix} R(0,0) & R(0,T) & R(0,2T) \\ R(T,0) & R(T,T) & R(T,2T) \\ R(2T,0) & R(2T,T) & R(2T,2T) \end{bmatrix} = \begin{bmatrix} 1 & 1 & 0 \\ 1 & 1 & 1 \\ 0 & 1 & 1 \end{bmatrix}$$

is not nonnegative definite. In fact, it has a negative determinant (-1).   ∎

The mean, autocorrelation, and autocovariance functions play a particularly important role in Gaussian processes. Recall that if $X(t)$ is Gaussian, the vector $\mathbf{X}(t) = [X(t_1), \ldots, X(t_n)]$ has the n-dimensional Gaussian density for any integer n and any $t_1, \ldots, t_n$. Thus, for a Gaussian random process $X(t)$, we need to know only $E\{X(t)\}$ and $\text{Cov}\{X(t), X(s)\}$ for all t and s in order to determine all of the n-dimensional distributions of $X(t)$. Therefore, the *mean* and *autocovariance* functions (or *mean* and *autocorrelation* functions) are sufficient to completely specify a Gaussian process. (See Section 2.7 for some further remarks on this point.)

2.3.2 Derivations of the Properties of Autocorrelation Functions

In this section, we show that (2.15)–(2.18) are satisfied by any autocorrelation function and, therefore, by any autocovariance function as well. This material is included for the reader who is interested in the detailed derivations of the key properties of autocorrelation and autocovariance functions. The material is not required for subsequent portions of the book, so it may be best to skip the section entirely during the first reading of this chapter.

The first step is to observe that (2.16) follows immediately from the definition of $R_X(t_1, t_2)$:

$$R_X(t_1, t_2) = E\{X(t_1)X(t_2)\} = E\{X(t_2)X(t_1)\} = R_X(t_2, t_1).$$

Next, notice that, for arbitrary choices of n; $t_1, \ldots, t_n$ in $\mathbb{T}$; and complex numbers $\alpha_1, \ldots, \alpha_n$, we have

$$0 \le E\left\{\left|\sum_{i=1}^{n} \alpha_i X(t_i)\right|^2\right\} = \sum_{i=1}^{n}\sum_{k=1}^{n} \alpha_i \alpha_k^* R_X(t_i, t_k),$$

which is just (2.18). Finally, observe that (2.15) is just a special case of (2.18): Let $n = 1, \alpha_1 = 1$, and $t_1 = t$.

The next step is to show that (2.18) implies (2.17). Recall that the latter is the inequality

$$|R_X(t_1, t_2)| \le \sqrt{R_X(t_1, t_1)R_X(t_2, t_2)}.$$

To derive this inequality from (2.18), it is helpful to begin by generalizing the approach used in the solution of Exercise 2–7. If $n = 2$ and α_1 and α_2 are real, (2.18) reduces to

$$\alpha_1^2 R_X(t_1, t_1) + \alpha_2^2 R_X(t_2, t_2) \ge -2\alpha_1\alpha_2 R_X(t_1, t_2). \tag{2.19}$$

Notice that the left-hand side of (2.19) does not depend on the sign of α_1 or of α_2. Hence, if λ_1 and λ_2 are arbitrary *nonnegative* numbers, then

$$\lambda_1^2 R_X(t_1, t_1) + \lambda_2^2 R_X(t_2, t_2) \ge 2\lambda_1\lambda_2 |R_X(t_1, t_2)|. \tag{2.20}$$

Inequality (2.20) can be derived from (2.19) by first letting $\alpha_1 = \lambda_1$ and $\alpha_2 = \lambda_2$ and then letting $\alpha_1 = \lambda_1$ and $\alpha_2 = -\lambda_2$. This step is a generalization of the approach used to obtain (2.13) and (2.14). Looking back at the derivation of those equations, we see that substituting $\alpha_1 = 1$ and $\alpha_2 = 1$ in (2.19) gives (2.13), while substituting $\alpha_1 = 1$ and $\alpha_2 = -1$ gives (2.14).

To proceed further, it is necessary to consider separately the cases in which $R_X(t_i, t_i) = 0$ for either $i = 1$ or $i = 2$ (or both). In a sense, such cases are degenerate, because they correspond to a random process for which the expected value of the instantaneous power is zero at time t_i. Nevertheless, we consider them for completeness. Both $R_X(t_1, t_1)$ and $R_X(t_2, t_2)$ are nonzero if and only if

$$R_X(t_1, t_1)R_X(t_2, t_2) > 0,$$

and this is the case that will be handled first. Recall that λ_1 and λ_2 can be any nonnegative numbers in (2.20). Assuming that $R_X(t_1, t_1)R_X(t_2, t_2) > 0$, we can substitute

$$\lambda_1^2 = \frac{R_X(t_2, t_2)}{2\sqrt{R_X(t_1, t_1)R_X(t_2, t_2)}}$$

and

$$\lambda_2^2 = \frac{R_X(t_1, t_1)}{2\sqrt{R_X(t_1, t_1)R_X(t_2, t_2)}}$$

into (2.20) to obtain (2.17). On the other hand, if at least one of these correlations is zero, then $R_X(t_1, t_1)R_X(t_2, t_2) = 0$, so the procedure will not work. However, if at least one of the two correlations is zero, we claim that the validity of (2.20) for all $\lambda_1 \geq 0$ and $\lambda_2 \geq 0$ implies $R_X(t_1, t_2) = 0$. If this claim is true, then (2.17) must hold, since the left-hand side is zero and the right-hand side is nonnegative.

It is particularly easy to establish that the claim is true if $R_X(t_i, t_i) = 0$ for *both* $i = 1$ and $i = 2$; we can just set $\lambda_1 = \lambda_2 = 1$ in (2.20) to deduce $|R_X(t_1, t_2)| \leq 0$. If $R_X(t_i, t_i) = 0$ for *only one* value of i, then we have to work a little harder. First, suppose that $R_X(t_1, t_1) = 0$ and $R_X(t_2, t_2) \neq 0$ (i.e., $i = 1$). Letting $\lambda_2 = 1$, we see that (2.20) reduces to

$$R_X(t_2, t_2) \geq 2\lambda_1|R_X(t_1, t_2)|.$$

Since this inequality must hold no matter how large λ_1 is, it must be that $R_X(t_1, t_2) = 0$; otherwise, as $\lambda_1 \to \infty$, the inequality will eventually be violated. For instance, if $R_X(t_1, t_2) \neq 0$ and we let

$$\lambda_1 = \frac{R_X(t_2, t_2)}{|R_X(t_1, t_2)|},$$

then the inequality becomes the assertion that $1 \geq 2$, which, of course, is false, so it must be that $R_X(t_1, t_2) = 0$, which establishes the claim for the case in which $R_X(t_1, t_1) = 0$ and $R_X(t_2, t_2) \neq 0$. The same argument with an interchange of appropriate 1's and 2's establishes the claim for the case in which $R_X(t_1, t_1) \neq 0$ and $R_X(t_2, t_2) = 0$.

We have shown that (2.15) and (2.17) follow from the nonnegative definiteness of the autocorrelation function. Although we obtained (2.16) directly from the definition of the autocorrelation function it also follows from the nonnegative definiteness of the autocorrelation function. We simply let $n = 2$, $\alpha_1^2 = -1$ (recall that α_1 can be complex), and $\alpha_2 = 1$, so that the *imaginary* part of the sum in (2.18) is $R_X(t_1, t_2) - R_X(t_2, t_1)$, which must be zero, since the sum must be a *real* number.

Thus, nonnegative definiteness implies all of the properties that we have considered. This is not surprising in view of the fact that nonnegative definiteness is a *necessary* and *sufficient* condition for a function $R(\cdot, \cdot)$ to be a valid autocorrelation. We already showed necessity when we derived (2.18). It turns out, however, that, given any nonnegative definite function $R(\cdot,\cdot)$ defined on the set of all (t_1, t_2) such that $t_1 \in \mathbb{T}$ and $t_2 \in \mathbb{T}$, there exists a random process $X(t), t \in \mathbb{T}$, that has autocorrelation function $R(\cdot,\cdot)$; that is,

$$R_X(t_1, t_2) = R(t_1, t_2)$$

for each t_1 and t_2. In fact, there always exists a *Gaussian* random process that has autocorrelation function $R(\cdot,\cdot)$. The proofs of these statements are beyond the scope and mathematical prerequisites of this book.

Finally, we should point out that (2.17) can be obtained as an immediate consequence of the *Schwarz inequality*, which states that, for any random variables Y_1 and Y_2,

$$E\{Y_1 Y_2\} \le \sqrt{E\{Y_1^2\} E\{Y_2^2\}}.$$

To derive (2.17) from the Schwarz inequality, simply let $Y_i = X(t_i)$ for both $i = 1$ and $i = 2$. One reason for deriving (2.17) from (2.20) rather than from the Schwarz inequality is that the derivation from (2.20) helps to bring out the importance of the nonnegative definiteness property: We have shown that the other properties can be deduced from the fact that a correlation function is nonnegative definite.

2.4   STATIONARY RANDOM PROCESSES

A random process $X(t), t \in \mathbb{T}$, is said to be *stationary* (or *strictly stationary*) if, for each n and each choice of $t_1, \ldots, t_n$ in $\mathbb{T}$,

$$F_{X,n}(x_1, \ldots, x_n; t_1, \ldots, t_n) = F_{X,n}(x_1, \ldots, x_n; t_1 + t_0, \ldots, t_n + t_0) \qquad (2.21)$$

for all t_0 in $\mathbb{T}$. In other words, $X(t)$ is stationary if all of its finite-dimensional distribution functions remain unchanged under all possible shifts in the time origin. The n-dimensional distributions, therefore, depend only on *relative* time (or time *differences*) for stationary random processes. This is a very important property for engineering applications, since the selection of a time origin is often a very artificial exercise. For many problems, it is important that the answer not depend on this arbitrary choice of a time origin. For obvious reasons, stationary processes are often called *shift-invariant* processes.

In (2.21), we must require that $t_i + t_0$ be in the index set $\mathbb{T}$ for all i; otherwise, the right-hand side of (2.21) is undefined. From this point on, then, assume that the index set is closed under addition (i.e., if t and s are in $\mathbb{T}$, then so is $t + s$). The most important examples of index sets that are closed under addition are the real line $\mathbb{R}$, the set of nonnegative real numbers $[0, \infty)$, the set of all integers $\mathbb{Z}$, and the set of nonnegative integers $\{0, 1, 2, \ldots\}$. It is convenient to make the additional assumption that the number 0 is in $\mathbb{T}$, which is true for all four of the preceding examples.

Example 2–6   A Satellite Beacon

Suppose a satellite transmits a tracking beacon to a ground station. The signal received at the ground station is a random process, even if the thermal noise in the receiver is ignored. A model for the received signal that is appropriate for some applications is the random process

$$X(t) = A \cos(2\pi f_0 t + \Theta).$$

The amplitude A of the signal is a random variable because of the random or unpredictable attenuation of the beacon signal as it propagates through the atmosphere. The phase angle Θ is also a random variable that may represent the randomness in the initial phase of the satellite's oscillator (as in Example 2–1) or the randomness in a phase shift that is introduced as the signal propagates through the atmosphere. In general, the random process $X(t)$ is not stationary unless conditions are placed on the distributions of A and Θ. For instance, suppose A and Θ are independent, $P(\Theta = 0) = P(\Theta = \pi) = \frac{1}{2}$, and A is a continuous random variable (e.g., Gaussian). Then $P[X(0) = 0] = 0$, but $P[X(\frac{1}{4}f_0) = 0] = 1$. Hence, even the one-dimensional distribution function depends on the parameter t, which violates (2.21).

However, for certain distributions of the random variable Θ, the random process $X(t)$ is stationary. For instance, if Θ is uniformly distributed on the interval $[0, 2\pi]$ and A is independent of Θ, then $X(t)$ is stationary. In practice, it is often true that Θ and A are independent and Θ is uniformly distributed on $[0, 2\pi]$. ∎

Example 2–7 A Nonstationary Process

Certain classes of engineering problems lend themselves to an obvious and natural choice for the time origin. For instance, if $X(t)$ is the number of telephone calls received at a switchboard, $t = 0$ represents the time at which we start counting calls. In this case, $X(t)$ is the number of calls received during the time interval $[0, t]$ (for $t \geq 0$). This is one example of a process that is inherently nonstationary according to its physical description. We would not expect $X(t_1)$ and $X(t_1 + \tau)$ to have the same distribution for $\tau > 0$, since we expect that, with high probability, there will be more calls received during $[0, t_1 + \tau]$ than during $[0, t_1]$. Hence, it is reasonable to conjecture that, for such a random process,

$$F_{X,1}(x_1; t_1) > F_{X,1}(x_1; t_1 + \tau), \quad \tau > 0.$$ ∎

For a second-order random process $X(t)$, a number of important consequences of stationarity are extremely useful in engineering. For the moment, think of t as being fixed and consider the random variable $X(t)$. The mean of this random variable depends only on the distribution function $F_{X,1}(\cdot; t)$. For example, if $F_{X,1}$ has derivative $F'_{X,1} = f_{X,1}$, then

$$\mu_X(t) = E\{X(t)\} = \int_{-\infty}^{\infty} u f_{X,1}(u; t)\, du.$$

If $X(t)$ is a discrete random variable that takes values in $\mathbb{Z}$, then

$$\mu_X(t) = E\{X(t)\} = \sum_{i=-\infty}^{\infty} i[F_{X,1}(i; t) - F_{X,1}(i - 1; t)].$$

If the process $X(t)$ is stationary, (2.21) implies that $F_{X,1}(x; t)$ does not depend on t, and thus, $\mu_X(t)$ does not depend on t. The conclusion is that *the mean of a stationary random process is a constant.* Whenever $\mu_X(t)$ does not depend on t, we denote its value by μ_X. This abuse of notation should cause no confusion, because it will be clear from the context whether the mean does or does not depend on the parameter t. If $\mu_X(t)$ depends on t, the process $X(t)$ is said to have a time-dependent or time-varying mean; otherwise, it is said to have a constant mean. For a process $X(t)$ with a time-varying mean, μ_X is a *function* with value $\mu_X(t)$ at time t. For a process $X(t)$ with a constant mean, μ_X is a real number.

If the process $X(t)$ is stationary, then (2.21) implies that $F_{X,2}(x_1, x_2; t + \tau, t)$ does not depend on t, although it may depend on τ. In other words, if $t_1 = t + \tau$ and $t_2 = t$, then $F_{X,2}(x_1, x_2; t_1, t_2)$ may depend on $t_1 - t_2$, but not on t_1 and t_2 individually. Since the autocorrelation and autocovariance functions depend only on the two-dimensional distribution functions, it follows that if $X(t)$ is stationary, then $R_X(t + \tau, t)$ and $C_X(t + \tau, t)$ do not depend on t.

The properties of the mean, autocorrelation, and autocovariance functions for a stationary, second-order random process are summarized as follows: If $X(t)$ is such a random process, then, for all t in $\mathbb{T}$,

$$\mu_X(t) = \mu_X(0), \tag{2.22}$$

$$R_X(t + \tau, t) = R_X(\tau, 0), \tag{2.23}$$

and

$$C_X(t + \tau, t) = C_X(\tau, 0) \tag{2.24}$$

for all τ in the index set $\mathbb{T}$. If (2.22) holds for all t, the mean is constant and is denoted by μ_X. Whenever $R_X(t + \tau, t)$ does not depend on t, but is instead a function of τ only, we write $R_X(\tau)$ in place of $R_X(t + \tau, t)$. Similarly, we denote $C_X(t + \tau, t)$ by $C_X(\tau)$ whenever (2.24) holds for all t and τ.

2.4.1   *Wide-Sense and Covariance Stationary Processes*

It may be that the functions μ_X, R_X, and C_X are shift invariant in the sense of equations (2.22)–(2.24), yet the process $X(t)$ is not stationary. Since those equations represent properties of a second-order process that are of interest in their own right, we define two forms of stationarity that are, in general, much weaker than (strict) stationarity as defined in (2.21). A second-order random process is said to be *wide-sense stationary* (*WSS*) if (2.22) and (2.23) hold for all t and τ; the process is said to be *covariance stationary* if (2.24) holds for all t and τ.

Letting $t_1 = t + \tau$ and $t_2 = t$ in (2.9), we find that

$$C_X(t + \tau, t) = R_X(t + \tau, t) - \mu_X(t + \tau)\mu_X(t).$$

From this relationship, we see that the right-hand side does not depend on t for a wide-sense stationary process, and, therefore, neither does the left-hand side. So $C_X(t + \tau, t)$ does not depend on t for a wide-sense stationary process. This establishes the following important fact:

A wide-sense stationary process is also covariance stationary.

However, there are important processes that are covariance stationary but *not* wide-sense stationary. One example arises in the investigation of a signal plus noise. If $X(t)$ is wide-sense stationary and $v(t)$ is a deterministic signal, then the random process

$$Y(t) = v(t) + X(t)$$

is covariance stationary, but it is wide-sense stationary only in the special case where $v(t)$ is a constant. The process $Y(t)$ might represent the voltage at some point in an electronic system, and this voltage consists of a signal component $v(t)$ plus a noise component $X(t)$. Only in the trivial situation in which the signal does not change with time is the resulting random process wide-sense stationary, and signals that do not change with time are of little interest in practice. This establishes another important fact:

A covariance stationary process need not be wide-sense stationary.

It is helpful to consider $Y(t) = v(t) + X(t)$ further to understand how such random processes can be covariance stationary but not wide-sense stationary. Recall that the covariance function is defined as

$$C_Y(t, s) = E\{[Y(t) - \mu_Y(t)][Y(s) - \mu_Y(s)]\},$$

and notice that the first step in computing this function is to subtract out the mean of the process $Y(t)$, which is given by

$$\mu_Y(t) = v(t) + \mu_X(t).$$

Although $\mu_X(t)$ is constant $[X(t)$ is WSS], the signal component $v(t)$ is not. So, it is the *mean* that keeps this particular process $Y(t)$ from being wide-sense stationary.

In determining covariance stationarity, we ignore the mean by subtracting it out. We then focus on the autocorrelation function for the new process that results when the mean is subtracted out of the original process. That is, we form the random process $W(t) = Y(t) - \mu_Y(t)$ and then evaluate the autocorrelation function

$$R_W(t, s) = E\{W(t)W(s)\}$$

for that process. If $Y(t) = v(t) + X(t)$, the resulting random process $W(t)$ is wide-sense stationary, as can be seen from the following: First, observe that $W(t)$ is a zero-mean process. Second, because $E\{Y(t)\} = v(t) + E\{X(t)\}$,

$$W(t) = Y(t) - \mu_Y(t) = Y(t) - [v(t) + \mu_X(t)] = X(t) - \mu_X(t),$$

so

$$R_W(t, s) = C_Y(t, s) = C_X(t, s) = R_X(t, s) - \mu_X(t)\mu_X(s).$$

Since $X(t)$ is wide-sense stationary, $\mu_X(t)$ does not depend on t, $\mu_X(s)$ does not depend on s, and $R_X(t, s)$ depends on t and s via the difference $t - s$ only. Thus, $R_W(t, s)$ also depends on t and s via the difference $t - s$ only. Because the mean of $W(t)$ is constant, $W(t)$ is wide-sense stationary, and because $R_W(t, s) = C_Y(t, s)$, $Y(t)$ is covariance stationary, which is what we promised to show.

If a random process is not wide-sense stationary, then either the mean or the auto-correlation function (or both) must depend on t. This requires the two-dimensional distribution function to depend on t, so a process that is not wide-sense stationary cannot be stationary. However, even if a random process $X(t)$ is wide-sense stationary, that does not guarantee, for example, that higher order moments $E\{[X(t)]^m\}$, $m > 2$, are constant with respect to t. Thus, wide-sense stationarity is a much weaker property than stationarity. We have already established that covariance stationarity does not even imply wide-sense stationarity, let alone stationarity. The relationships between wide-sense stationarity and stationarity for second-order random processes are summarized as follows:

A stationary process is also wide-sense stationary.
A wide-sense stationary process need not be stationary.

For a wide-sense stationary process $X(t)$, (2.15)–(2.17) become

$$R_X(0) \geq 0, \tag{2.25}$$

$$R_X(\tau) = R_X(-\tau), \tag{2.26}$$

and

$$|R_X(\tau)| \leq R_X(0), \tag{2.27}$$

for any τ.

Notice that the inequality $|R_X(t,s)| \leq [R_X(t,t) + R_X(s,s)]/2$, which is just (2.12), and the inequality $|R_X(t,s)| \leq \sqrt{R_X(t,t)R_X(s,s)}$, which is (2.17), are equivalent if $X(t)$ is wide-sense stationary.

Example 2–8   Sinusoidal Signal with Random Amplitude and Phase

It was pointed out in Example 2–6 that the process $X(t) = A\cos(2\pi f_0 t + \Theta)$ is not stationary in general. In fact, the process may not even be wide-sense stationary, as we illustrate next. Suppose that A and Θ are independent and $E\{A^2\} < \infty$. Then

$$E\{X(t_1)X(t_2)\} = \frac{1}{2}E\{A^2\}\{\cos[2\pi f_0(t_1 - t_2)] + E\{\cos 2\Theta\}\cos[2\pi f_0(t_1 + t_2)]$$
$$- E\{\sin 2\Theta\}\sin[2\pi f_0(t_1 + t_2)]\},$$

which, in general, depends not only on $t_1 - t_2$, but also on $t_1 + t_2$. An important special case is when $E\{\cos 2\Theta\} = E\{\sin 2\Theta\} = 0$, in which case $R_X(t,t) = \frac{1}{2}E\{A^2\}$ and

$$R_X(t + \tau, t) = \frac{1}{2}E\{A^2\}\cos(2\pi f_0 \tau),$$

which depends on τ only (not on t). If, in addition, $E\{\cos\Theta\} = E\{\sin\Theta\} = 0$, then $\mu_X(t) = 0$ for each t, so that $X(t)$ is wide-sense stationary.

Thus, the process $X(t) = A\cos(2\pi f_0 t + \Theta)$ is a second-order, wide-sense-stationary random process if A and Θ are independent, $E\{A^2\} < \infty$, and

$$E\{\cos\Theta\} = E\{\sin\Theta\} = E\{\cos 2\Theta\} = E\{\sin 2\Theta\} = 0.$$

The latter condition holds if, for instance, Θ is uniformly distributed on $[0, 2\pi]$.   ■

Exercise 2–11. Show that Example 2–8 can be used to construct a random process that is wide-sense stationary, but not stationary.

Solution. Let $A = 1$ in Example 2–8, so that $X(t) = \cos(2\pi f_0 t + \Theta)$. Let Θ take on each of the values $0, \pi/2, \pi$, and $3\pi/2$ with probability $\frac{1}{4}$. It is easy to show that

$$E\{\cos\Theta\} = E\{\sin\Theta\} = E\{\cos 2\Theta\} = E\{\sin 2\Theta\} = 0,$$

so that, according to Example 2–8, the resulting random process is wide-sense stationary. There are several ways to show that $X(t)$ is not stationary, such as observing that $P[X(0) = 0] = \frac{1}{2}$ and $P[X(0) = +1] = P[X(0) = -1] = \frac{1}{4}$, but if $t_0 = (8f_0)^{-1}$, the random variable $X(t_0)$ takes on each of the two values $+1/\sqrt{2}$ and $-1/\sqrt{2}$ with probability $\frac{1}{2}$. Clearly, $X(0)$ and $X(t_0)$ do not have the same distribution.   ■

Example 2–9   A Simple Nonstationary Process

Let $X(t) = Y + tZ$, where Y and Z are random variables with finite second moments. The random process $X(t)$ is therefore a second-order random process. It is clear from the beginning that, unless Z is a trivial random variable (i.e., $P(Z = 0) = 1$), the process $X(t)$ is not stationary. To prove that it is not, it suffices to prove that $X(t)$ is not wide-sense stationary. Toward that end, let μ_Y and μ_Z be the mean values of Y and Z, respectively. Straightforward evaluation then shows that the autocorrelation function is

$$R_X(t + \tau, t) = E\{Y^2\} + (2t + \tau)E\{YZ\} + t(t + \tau)E\{Z^2\}$$

and the mean function is $\mu_X(t) = \mu_Y + \mu_Z t$. The process $X(t)$ is not, in general, wide-sense stationary, because both the mean and the autocorrelation depend on t, rather than having a constant mean and an autocorrelation that depends on the time difference τ alone. Even if $\mu_Z = 0$, so that $\mu_X(t) = \mu_Y$ (a constant) for all t, the process is still not necessarily wide-sense stationary. For instance, if $\mu_Z = 0$ and the random variables Y and Z are uncorrelated, then

$$R_X(t + \tau, t) = E\{Y^2\} + t(t + \tau)\,\mathrm{Var}\{Z\},$$

which depends on t.   ■

A Gaussian random process is completely specified by its mean and autocorrelation functions. The reason for this is that if $X(t)$ is a Gaussian random process, the joint distribution function for $X(t_1), X(t_2), \ldots, X(t_n)$ depends only on the means and covariances of these n random variables. The means are given by

$$\mu_X(t_i) = E\{X(t_i)\}, \quad 1 \le i \le n,$$

and the covariances are given by

$$C_X(t_i, t_j) = \mathrm{Cov}\{X(t_i), X(t_j)\}, \quad 1 \le i \le n, 1 \le j \le n.$$

Thus, the n-dimensional distributions for a Gaussian process are independent of the time origin if and only if the mean of the process is constant and its autocorrelation function does not depend on the time origin. (See Section 2.7.) From the relationship

$$C_X(t, s) = R_X(t, s) - \mu_X(t)\mu_X(s),$$

it follows that a necessary and sufficient condition for the n-dimensional distributions for a Gaussian process to be independent of the time origin is that the mean and autocorrelation do not depend on the time origin.

We conclude that a Gaussian random process $X(t)$ satisfies

$$F_{X,n}(x_1, \ldots, x_n; t_1, \ldots, t_n) = F_{X,n}(x_1, \ldots, x_n; t_1 + t_0, \ldots, t_n + t_0)$$

for all $t_0, t_1, \ldots, t_n$ in $\mathbb{T}$ if and only if it satisfies $\mu_X(t) = \mu_X(0)$ for all t in $\mathbb{T}$ and, for each τ, $R_X(t + \tau, t) = R_X(\tau, 0)$ for all t in $\mathbb{T}$. This is equivalent to the statement that a Gaussian random process is stationary whenever it is wide-sense stationary.

The relationship between stationary and nonstationary second-order random processes and the special role of Gaussian processes are summarized as follows:

A stationary random process is always wide-sense stationary. A wide-sense stationary process need not be stationary. However, for a Gaussian random process, stationarity and wide-sense stationarity are equivalent.

2.4.2 *Examples of Autocorrelation Functions for Wide-Sense Stationary Processes*

In this section, we give some common examples of autocorrelation functions that depend on time differences only. Some of these autocorrelation functions arise in subsequent sections and in homework problems as the autocorrelation functions for certain random processes. For now, we simply introduce them without connection to specific processes. The parameters α and β that appear in the autocorrelation functions are positive in each case. The reader should verify that each of the autocorrelation functions presented in Table 2–1 satisfies (2.25)–(2.27).

From these examples, it is possible to obtain the autocorrelation functions for some other random processes by using certain analytical results that give the autocorrelation function for a random process defined in terms of one or more other random processes. For example, suppose that $Y(t) = cX(t)$ for some constant c. The autocorrelation function for $Y(t)$ is given by

$$R_Y(t, s) = E\{Y(t)Y(s)\} = E\{[cX(t)][cX(s)]\} = c^2 E\{X(t)X(s)\} = c^2 R_X(t, s).$$

Clearly, if $X(t)$ is wide-sense stationary, then so is $Y(t)$, and the identity becomes

$$R_Y(\tau) = c^2 R_X(\tau).$$

Other examples will be given after we introduce the notion of the crosscorrelation function for two random processes.

TABLE 2–1   Some Examples of Autocorrelation Functions

1. $R(\tau) = \beta \exp(-\alpha|\tau|)$ (illustrated in Figure 2–11(a))

2. $R(\tau) = \begin{cases} \beta(T - |\tau|)/T, & -T \leq \tau \leq T \\ 0, & \text{otherwise} \end{cases}$ (illustrated in Figure 2–11(b))

3. $R(\tau) = \beta \exp(-\alpha|\tau|) \cos(\omega_0 \tau)$

4. $R(\tau) = 2W\{\sin(2\pi W \tau)/2\pi W \tau\}$

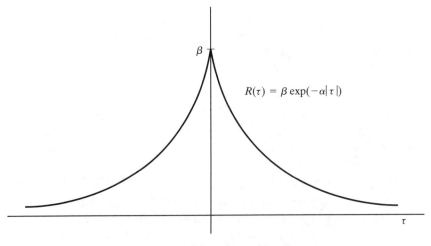

FIGURE 2–11(a)   The exponential autocorrelation function.

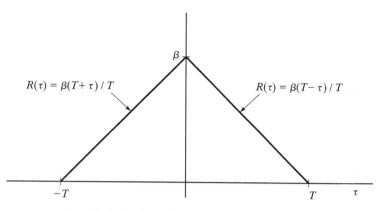

FIGURE 2–11(b)   The triangular autocorrelation function.

2.5   CROSSCORRELATION AND CROSSCOVARIANCE FUNCTIONS

Let $X(t), t \in \mathbb{T}$, and $Y(t), t \in \mathbb{T}$, be two second-order random processes defined on the same probability space. The *crosscorrelation function* $R_{X,Y}$ for these two random processes is defined by

$$R_{X,Y}(t, s) = E\{X(t)Y(s)\}$$

for all t and s. The crosscorrelation function satisfies

$$R_{X,X}(t, s) = R_X(t, s), \tag{2.28}$$
$$R_{X,Y}(t, s) = R_{Y,X}(s, t), \tag{2.29}$$

and

$$\left| R_{X,Y}(t, s) \right| \leq \sqrt{R_X(t, t) R_Y(s, s)} \tag{2.30}$$

for all t and s. The *crosscovariance function* $C_{X,Y}$ is defined by

$$C_{X,Y}(t, s) = E\{[X(t) - \mu_X(t)][Y(s) - \mu_Y(s)]\}$$

for all t and s. Notice that the crosscovariance function and crosscorrelation function are related by

$$C_{X,Y}(t, s) = R_{X,Y}(t, s) - \mu_X(t)\mu_Y(s).$$

The two second-order random processes $X(t), t \in \mathbb{T}$, and $Y(t), t \in \mathbb{T}$, are said to be *uncorrelated* if, for all t in $\mathbb{T}$ and all s in $\mathbb{T}$,

$$R_{X,Y}(t, s) = \mu_X(t)\mu_Y(s).$$

Observe from the relationship between the crosscovariance and crosscorrelation functions that $X(t)$ and $Y(t)$ are uncorrelated if and only if $C_{X,Y}(t, s) = 0$ for all t and s.

The processes $X(t), t \in \mathbb{T}$, and $Y(t), t \in \mathbb{T}$, are said to be *independent random processes* if, for each positive integer n and each choice of $t_1, t_2, \ldots, t_n, s_1, s_2, \ldots, s_n$ in $\mathbb{T}$, the two random vectors $[X(t_1), X(t_2), \ldots, X(t_n)]$ and $[Y(s_1), Y(s_2), \ldots, Y(s_n)]$ are (statistically) independent. Two random vectors $\mathbf{X}$ and $\mathbf{Y}$ are independent if and only if their joint distribution function factors; that is, **X** and **Y** are independent if and only if $F_{X,Y}(x, y) = F_X(x)F_Y(y)$ for each pair **x** and **y**.

If $X(t)$ and $Y(t)$ are independent random processes, then, for each choice of t_0 and s_0 in $\mathbb{T}$, $X(t_0)$ and $Y(s_0)$ are independent random variables. Because independent random variables are uncorrelated, it follows that independent random processes are also uncorrelated. In general, however, $X(t), t \in \mathbb{T}$, and $Y(t), t \in \mathbb{T}$, can be uncorrelated random processes without being independent random processes, as is demonstrated by the next example.

Example 2–10   A Pair of Uncorrelated Random Processes

Let Z be a zero-mean Gaussian random variable, and define the random processes $X(t), -\infty < t < \infty$, and $Y(t), -\infty < t < \infty$, by

$$X(t) = Zt$$

for all t and

$$Y(t) = Z^2 t$$

for all t. First, observe that these random processes are not independent. For instance, $Y(1) = [X(1)]^2$, so that $X(1)$ and $Y(1)$ are not even independent. More generally, for any $t_0 \neq 0$, $Y(t) = [X(t_0)/t_0]^2 t$ for all t. Thus, if we observe the process $X(t)$ at any nonzero time, we can determine the value of the process $Y(t)$ for all t. This is a very strong form of statistical dependence!

Next, we show that these random processes are uncorrelated. Because Z is a zero-mean Gaussian random variable, its density function is symmetric; that is, $f_Z(z) = f_Z(-z)$ for all z. Thus, all odd-order moments of Z are zero. In particular, $E\{Z\} = 0$, so $\mu_X(t) = 0$ for all t, and $E\{Z^3\} = 0$, so

$$R_{X,Y}(t, s) = E\{X(t)Y(s)\} = E\{(Zt)(Z^2 s)\} = tsE\{Z^3\} = 0$$

for all t and s. Recall that the two processes are uncorrelated if, for all t and s,

$$R_{X,Y}(t, s) = \mu_X(t)\mu_Y(s).$$

Clearly, this relationship is satisfied, because each side of the equation is equal to zero. Hence, $X(t)$ and $Y(t)$ are uncorrelated random processes that are not independent.   ■

The random processes $X(t), t \in \mathbb{T}$, and $Y(t), t \in \mathbb{T}$, are said to be *jointly Gaussian random processes* if, for each positive integer n and each choice of $t_1, t_2, \ldots, t_n, s_1, s_2, \ldots, s_n$ in $\mathbb{T}$, the two random vectors $[X(t_1), X(t_2), \ldots, X(t_n)]$ and $[Y(s_1), Y(s_2), \ldots, Y(s_n)]$ are jointly Gaussian. That is, the single random vector

$$[X(t_1), X(t_2), \ldots, X(t_n), Y(s_1), Y(s_2), \ldots, Y(s_n)]$$

has a distribution governed by the $2n$-dimensional Gaussian density. Because uncorrelatedness and independence are equivalent for jointly Gaussian *random variables*, jointly Gaussian *random processes* that are uncorrelated are also independent. Notice that, although the random process $X(t)$ in Example 2–10 is Gaussian, the random process $Y(t)$ is not, so the two processes are certainly not jointly Gaussian. The relationship between uncorrelatedness and independence for two random processes is summarized as follows:

> *Independent random processes are uncorrelated. In general, uncorrelated random processes need not be independent. However, jointly Gaussian random processes that are uncorrelated are also independent.*

The random processes $X(t)$ and $Y(t)$ are said to be *jointly wide-sense stationary* if $X(t)$ is wide-sense stationary, $Y(t)$ is wide-sense stationary, and $R_{X,Y}(t, s)$ depends on $t - s$ only. Clearly, joint wide-sense stationarity of a pair of random processes is, in general, a stronger condition than wide-sense stationarity of each of the processes. The condition that $R_{X,Y}(t, s)$ depend on $t - s$ only does not arise in determining the wide-sense stationarity of the individual random processes. This condition can be expressed as $R_{X,Y}(t, s) = R_{X,Y}(t - s, 0)$, for all t and s in $\mathbb{T}$. If the processes $X(t)$ and $Y(t)$ are jointly wide-sense stationary, we usually write $R_{X,Y}(t - s)$ in place of $R_{X,Y}(t, s)$ and $R_{X,Y}(\tau)$ in place of $R_{X,Y}(t + \tau, t)$.

Recall that if $X(t), t \in \mathbb{T}$, and $Y(t), t \in \mathbb{T}$, are uncorrelated random processes, then $C_{X,Y}(t, s) = 0$ for all t and s, so that $R_{X,Y}(t, s) = \mu_X(t)\mu_Y(s)$ for all t and s. If these random processes are each wide-sense stationary, then $\mu_X(t)$ and $\mu_Y(s)$ are the constants μ_X and μ_Y, respectively. So, if $X(t)$ and $Y(t)$ are uncorrelated random processes that are each wide-sense stationary, the product $\mu_X(t)\mu_Y(s)$ is just the constant $\mu_X\mu_Y$ for all t and s. Hence, $R_{X,Y}(t, s) = \mu_X\mu_Y$ does not depend on t or s at all. Thus, *uncorrelated* random processes that are *individually* wide-sense stationary are also *jointly* wide-sense stationary. In general, however, two wide-sense stationary processes are not necessarily jointly wide-sense stationary.

We can now return to the topic of the autocorrelation function for a random process defined in terms of two (or more) other random processes. For the remainder of this section, let $X(t), t \in \mathbb{T}$, and $Y(t), t \in \mathbb{T}$, be random processes defined on the same probability space. The autocorrelation function for the random process defined by $Z(t) = X(t) + Y(t)$ is given by

$$R_Z(t, s) = E\{[X(t) + Y(t)][X(s) + Y(s)]\} =$$
$$R_X(t, s) + R_{X,Y}(t, s) + R_{Y,X}(t, s) + R_Y(t, s). \quad (2.31)$$

If each random process has *zero mean* and the two random processes are *uncorrelated*, then (2.31) reduces to

$$R_Z(t, s) = R_X(t, s) + R_Y(t, s). \quad (2.32)$$

Thus, for zero-mean, uncorrelated random processes, the autocorrelation function for the sum of two random processes is the sum of the individual autocorrelation functions. It should be clear that this relationship extends to the sum of any finite number of zero-mean, uncorrelated random processes. If, in addition, *each* of the two random processes is *wide-sense stationary* (note that this plus the fact that they are uncorrelated implies that they are jointly wide-sense stationary), then (2.32) becomes $R_Z(t - s) = R_X(t - s) + R_Y(t - s)$, or simply,

$$R_Z(\tau) = R_X(\tau) + R_Y(\tau). \quad (2.33)$$

It follows that the sum of any two autocorrelation functions is a valid autocorrelation function.

Next, we consider the product of two random processes: $Z(t) = X(t)Y(t)$ for $t \in \mathbb{T}$. The autocorrelation function for the random process $Z(t)$ is given by

$$R_Z(u, s) = E\{X(u)Y(u)X(s)Y(s)\}.$$

If $X(t), t \in \mathbb{T}$, and $Y(t), t \in \mathbb{T}$, are independent random processes, then $X(u)X(s)$ and $Y(u)Y(s)$ are independent for each choice of u and s in $\mathbb{T}$. It follows that

$$E\{X(u)Y(u)X(s)Y(s)\} = E\{X(u)X(s)\}E\{Y(u)Y(s)\} = R_X(u, s)R_Y(u, s)$$

for all u and s, and therefore,

$$R_Z(u, s) = R_X(u, s)R_Y(u, s) \tag{2.34}$$

for all u and s. That is, for independent random processes, the autocorrelation function for the product of two random processes is the product of the individual autocorrelation functions. This relationship also extends easily to the product of any finite number of independent random processes. The reader should verify that it is not sufficient that the random processes be uncorrelated. A stronger condition, such as independence, is required in order that $E\{X(u)Y(u)X(s)Y(s)\}$ can be written as the product of $E\{X(u)X(s)\}$ and $E\{Y(u)Y(s)\}$. Note that $E\{X(u)Y(u)X(s)Y(s)\}$ is a fourth-order moment, and uncorrelatedness tells us something about second-order moments only. If the random processes are independent and each is wide-sense stationary, (2.34) can be written as $R_Z(u - s) = R_X(u - s)R_Y(u - s)$, or simply,

$$R_Z(\tau) = R_X(\tau)R_Y(\tau). \tag{2.35}$$

It follows that the product of any two autocorrelation functions is a valid autocorrelation function.

Example 2–11   Amplitude Modulation

Suppose that the wide-sense stationary random process $X(t)$ is amplitude modulated onto an RF carrier to produce the signal

$$Z(t) = \sqrt{2}X(t)\cos(2\pi f_c t + \Theta),$$

where Θ is uniformly distributed on $[0, 2\pi]$ and is independent of $X(t)$. If we let

$$Y(t) = \sqrt{2}\cos(2\pi f_c t + \Theta),$$

we see from Exercise 2–8 that

$$R_Y(\tau) = \cos(2\pi f_c \tau).$$

If Θ is independent of $X(t)$, any deterministic function of Θ is also independent of $X(t)$. Also, $X(t)$ and $Y(t)$ are each wide-sense stationary, so that (2.35) applies and establishes that the autocorrelation function for $Z(t)$ is

$$R_Z(\tau) = R_X(\tau)\cos(2\pi f_c \tau). \qquad\blacksquare$$

2.6   WHITE NOISE

In the previous sections, we have considered only random processes that satisfy $E[X(t)]^2 < \infty$ for all t. This is simply a requirement that the process have finite power, which we expect to be true of all processes that arise in practical problems. Such processes are what we have referred to as *second-order* random processes. All of the processes that we will be concerned with are second-order processes, with one very important exception: white noise.

White noise is a mathematical idealization of the thermal noise process discussed in Example 2–2. In fact, as described there, the random process that models thermal noise has infinite power, since it has power equal to $4kTRB$ in each frequency band of B Hz. If we integrate this power over all frequencies, we see that the random process has infinite power. However, two important points should be considered: First, the model discussed in the example is not valid for arbitrarily large frequencies; second, no physical system has infinite bandwidth, and we can observe the thermal noise process only at the output of some physical system. Even if the input process is modeled as white noise, the output of a physical system will not have a fixed amount of power in each part of the frequency band. In fact, it will decrease as the frequency increases in a way that gives finite power. Thus, it can be *assumed* (and it is mathematically convenient to do so) that thermal noise is as described in Example 2–2 for all values of f_0, even though this leads to an assumption of infinite power. In most situations, it makes no difference whether we restrict f_0 to be less than 10^{11} Hz or 10^{12} Hz or allow all finite values of f_0, because noise at frequencies greater than about 10^{11} Hz will have no effect on the output of most electronic systems.

The autocorrelation function of a stationary white-noise process is taken to be a delta function centered at the origin; that is, $E\{X(t + \tau)X(t)\} = c\delta(\tau)$ for some positive constant c. As we will see later, this actually follows from the earlier assumption that the noise power is the same in all frequency bands of a given bandwidth.

We conclude that white noise is *not* a second-order process, since it has an infinite second moment. Strictly speaking, $E\{[X(t)]^2\}$ is not even defined, according to the delta-function autocorrelation model. The reason for wanting to employ the white-noise model we have described is that it leads to a considerable simplification of the analysis of thermal noise in linear systems.

2.7   STATIONARY GAUSSIAN RANDOM PROCESSES

We begin with the following fact about jointly Gaussian random variables.

Fact: The distribution of a collection of n jointly Gaussian random variables $X_1, X_2, \ldots, X_n$ depends only on the mean values $\mu_i = E\{X_i\}$ and the covariances $\text{Cov}\{X_i, X_j\}$ for the n random variables.

This fact follows from the discussion of Gaussian random variables in Section 1.2.3, where it is pointed out that if Λ is the $n \times n$ matrix that has

$$\Lambda_{i,j} = E\{(X_i - \mu_i)(X_j - \mu_j)\} = \text{Cov}\{X_i, X_j\} \qquad (2.36)$$

as the element in the ith row and jth column, and if μ_i is the mean of X_i ($1 \leq i \leq n$), then the joint density function for $\mathbf{X} = (X_1, X_2, \ldots, X_n)$ is given by

$$f_{\mathbf{X},n}(\boldsymbol{u}) = (2\pi)^{-n/2} |\det(\Lambda)|^{-1/2} \exp\left\{-\tfrac{1}{2}(\boldsymbol{u} - \boldsymbol{\mu})\Lambda^{-1}(\boldsymbol{u} - \boldsymbol{\mu})^T\right\}, \qquad (2.37)$$

where $\det(\Lambda)$ is the determinant of the matrix Λ, Λ^{-1} is the inverse of Λ, $\mathbf{u} = (u_1, u_2, \ldots, u_n)$, $\boldsymbol{\mu} = (\mu_1, \mu_2, \ldots, \mu_n)$, and $(\mathbf{u} - \boldsymbol{\mu})^T$ is the transpose of the vector $(\mathbf{u} - \boldsymbol{\mu})$. The matrix Λ is the covariance matrix for the random vector $\mathbf{X}$, and the vector $\boldsymbol{\mu}$

is the mean vector for **X**. We see from (2.37) that $\boldsymbol{\mu}$ and Λ completely characterize the density function; furthermore, $\boldsymbol{\mu}$ and Λ depend only on the means and covariances of the random variables $X_1, X_2, \ldots, X_n$.

To apply this fact to a Gaussian random process $X(t)$, let $X_i = X(t_i)$ for each i, and let $\mathbf{X} = (X_1, X_2, \ldots, X_n)$. Then

$$\mu_i = E\{X(t_i)\} = \mu_X(t_i) \tag{2.38}$$

and

$$\Lambda_{i,j} = \mathrm{Cov}\{X(t_i), X(t_j)\} = C_X(t_i, t_j). \tag{2.39}$$

If the Gaussian random process $X(t)$ is wide-sense stationary, the mean $\mu_X(t_i)$ does not depend on t_i, and the autocovariance $C_X(t_i, t_j)$ depends on the time differences $t_i - t_j$ only. Consequently, the n-dimensional density function for the wide-sense stationary Gaussian random process $X(t)$ depends only on the differences between the sample times. In fact, this density is given in terms of the density of (2.37) by

$$f_{X,n}(u_1, u_2, \ldots, u_n; t_1, t_2, \ldots, t_n) = f_X(\boldsymbol{u}). \tag{2.40}$$

Since all this is true for all choices of the integer n, we conclude that a wide-sense stationary Gaussian random process is also (strictly) stationary.

PROBLEMS

2.1 Let $g(t)$ denote the periodic "sawtooth" signal defined in the interval $\left[\dfrac{-T}{2}, \dfrac{T}{2}\right]$ by

$$g(t) = 1 - \frac{4t}{T} \quad \text{for } 0 \le t \le \frac{T}{2}$$

and

$$g(t) = 1 + \frac{4t}{T} \quad \text{for } \frac{-T}{2} \le t \le 0.$$

For other intervals, $g(t)$ is defined in a way to make it periodic with period T; that is, for $(2n-1)\dfrac{T}{2} \le t \le (2n+1)\dfrac{T}{2}$, $g(t) = g(t - nT)$. A random process is given by $X(t) = g(t - V)$, where V is a random variable.

   (a) Sketch some typical sample functions for the random process $X(t)$ if V is uniformly distributed on the interval $[0, T]$.

   (b) Sketch *all* possible sample functions if V is distributed according to $P(V = mT/4) = \frac{1}{4}$ for $m = 0, 1, 2$, and 3.

2.2 Consider a checkout line at a local grocery store. Let $N(t)$ denote the number of customers (including the one being served) waiting in line at time t. Sketch three typical sample functions for this random process. What is the primary difference between these sample functions and that of Figure 2–5?

2.3 Consider a packet communication network with K communication terminals, each of which can simultaneously transmit and receive. The network has the capability of allowing any subset of these terminals to transmit packets at a given time (i.e., packet transmissions can overlap). Each packet requires T seconds of transmission time. Let $N(t)$ denote the number of packets being transmitted at time t. Sketch three typical sample functions for each of the following situations:

(a) Slotted transmission is used, and the clocks at all terminals are completely synchronized. That is, time is divided into T-second slots, each transmission must start at the beginning of a slot and stop at the end of the same slot, and each terminal "knows" precisely when the slots begin and end.

(b) The terminals are operated asynchronously and the transmissions are not slotted. In this case, a terminal begins its transmission anytime it has a packet ready to transmit.

2.4 What are the *two* major differences between the sample functions of Problem 2.3(b) and those of Problem 2.2? (*Hint*: Consider the maximum value and the duration of "service".)

2.5 Determine the classification of the random processes in each of the situations that follow. In each case, indicate whether the process is a continuous- or discrete-time process and whether it is a continuous- or discrete-amplitude process.

(a) A manufacturing process begins at time 0, and we are interested in the number of defects that have occurred up to time t for all positive values of t.

(b) A computation is carried out in a sequence of steps in a special-purpose digital computer, and the content of a particular shift register is converted to decimal form and recorded at the end of each clock cycle.

(c) A continuous-time signal is sampled and quantized every T_0 seconds, and we are interested in the quantization error at the sampling times.

(d) A continuous-time signal is sampled and quantized every T_0 seconds, and we are interested in the quantized value of the signal at each sampling time.

(e) Because of interference and thermal noise in an analog FM receiver, the demodulated audio signal differs from the transmitted audio signal. We are interested in the error signal (the difference between the transmitted and demodulated audio signals).

2.6 In this problem, we make a minor modification to Example 2–4. The output process is $Y(t) = X(t) + X(t - 1) + X(t - 2)$ for $t \geq 2$. Find the one-dimensional density function for the random process $Y(t), t \geq 2$. Also, find $P[Y(t) = 3, Y(t + 1) = 2]$ for $t \geq 2$.

2.7 Suppose X is a random variable that is uniformly distributed on $[0, 1]$. The random process $Y(t), t > 0$, is defined by $Y(t) = \exp\{-Xt\}$. Find the one-dimensional distribution function $F_{Y,1}(u; t)$ for the random process $Y(t)$. Find the one-dimensional density function $f_{Y,1}(u; t)$.

2.8 The random process $X(t), t \in \mathbb{R}$, is wide-sense stationary with mean μ_X and autocorrelation function $R_X(\tau)$. Suppose that for $|\tau| > \tau_0$, the random variables $X(t_0)$ and $X(t_0 + \tau)$ are uncorrelated for all t_0. What is the value of $R_X(\tau)$ for $\tau > \tau_0$? In particular, what is the value of $\lim_{\tau \to \infty} R_X(\tau)$?

2.9 Suppose that $X(t)$ is a zero-mean, wide-sense stationary, continuous-time Gaussian random process with autocorrelation function $R_X(\tau)$. Let the random process $Y(t)$ be given by $Y(t) = c_1 X(t) + c_2 X(t - T)$. Find the probability that $Y(t_0)$ is greater than some threshold γ. Express your answer in terms of the standard Gaussian distribution function Φ, the autocorrelation function R_X, and the parameters c_1, c_2, γ, and T.

2.10 The wide-sense stationary random process $X(t), t \in \mathbb{R}$, is a Gaussian random process with zero mean and autocorrelation function $R_X(\tau)$. Find the one-dimensional distribution function for the random process $Y(t) = \Phi\big(X(t)/\sqrt{R_X(0)}\big)$ for $-\infty < t < \infty$. Find the one-dimensional density function for $Y(t)$. Is the process $Y(t)$ stationary?

2.11 Consider the function R given by $R(\tau) = 1, -T \leq \tau \leq T$, and $R(\tau) = 0$ otherwise. Does this function satisfy (2.25)–(2.27)? Is it a valid autocorrelation function? Prove your answer to each question.

2.12 Does the function $R(\tau) = \beta \exp(-\alpha\tau^2)$ satisfy (2.25)–(2.27) if β and α are positive real numbers? Prove your answer.

2.13 State whether $R(\tau) = \exp\{-(\tau - 1)^2\}$ does or does not satisfy (2.25)–(2.27), and prove your answer.

2.14 A random process $X(t)$ has a constant mean and an autocorrelation function

$$R_X(t, s) = \cos(\omega_0 t) \cos(\omega_0 s) + \sin(\omega_0 t) \sin(\omega_0 s).$$

Is the random process wide-sense stationary? Prove your answer.

2.15 Let $X(t) = A \cos(2\pi \Lambda t + \Theta), -\infty < t < \infty$. This is a sinusoidal signal with random amplitude, frequency, and phase. Suppose that A, Λ, and Θ are mutually independent random variables. Suppose also that A is a nonnegative random variable with density $f_A(a) = 0.1 \exp(-a/10), a > 0, \Lambda$ is uniformly distributed on the interval $[-W, W]$, and Θ is uniformly distributed on $[0, 2\pi]$. Find the mean and autocorrelation functions for the random process $X(t)$. Is $X(t)$ wide-sense stationary?

2.16 Suppose $X(t)$ and $Y(t)$ are zero-mean, wide-sense stationary, continuous-time random processes. If $X(t)$ and $Y(t)$ are independent, find the autocorrelation function for $Z(t)$ in terms of the autocorrelation functions for $X(t)$ and $Y(t)$ in each of the cases that follow. In each case, determine whether the random process $Z(t)$ is wide-sense stationary.
(a) $Z(t) = cX(t)Y(t) + d$, where c and d are deterministic constants
(b) $Z(t) = X(t) \cos(\omega_0 t) + Y(t) \sin(\omega_0 t)$

2.17 Consider the random processes

$$X(t) = \alpha_1 \cos(2\pi f_0 t + \Theta)$$

and

$$Y(t) = \alpha_2 \sin(2\pi f_0 t + \Theta),$$

where α_1 and α_2 are deterministic constants and Θ is uniformly distributed on $[0, 2\pi]$.
(a) Is each of these random processes wide-sense stationary?
(b) Find the crosscorrelation function for the two random processes.
(c) Are these random processes jointly wide-sense stationary?
(d) Are the two random processes uncorrelated?
(e) Are the two random processes independent?

2.18 Suppose V is a zero-mean Gaussian random variable, and define the random processes $X(t) = Vt$ and $Y(t) = V^2 t$ for $-\infty < t < \infty$.
(a) Find the crosscorrelation function for these two random processes.
(b) Are these random processes jointly wide-sense stationary?

2.19 A wide-sense stationary Gaussian random process $X(t), -\infty < t < \infty$, has autocorrelation function given by $R_X(\tau) = \beta(T - |\tau|)/T$ for $|\tau| \leq T$ and $R_X(\tau) = 0$ otherwise, as illustrated in Figure 2–11(b). The mean of this random process is zero.
(a) Show that $\beta \geq 0$, and show that $\beta = 0$ only if the process is trivial in some sense. Define in what sense it is trivial if $\beta = 0$.
(b) Find the one-dimensional distribution function for this random process.
(c) Find the two dimensional density function $f_{X,2}(u_1, u_2; t_1, t_2)$ for $t_1 = T$ and $t_2 = 2T$.
(d) Repeat **(c)** for $t_1 = T/2$ and $t_2 = T$.

(e) Find the probability that $X(0) + X(T/2) + X(T)$ exceeds 10.

In **(c)** and **(d)**, your answers should be expressed in terms of the parameters β and T only (not t_1 and t_2). In **(b)**–**(e)**, simplify your answers as much as possible.

2.20 A wide-sense stationary random process $X(t)$ has autocorrelation function

$$R_X(\tau) = c_1 \exp(-c_2|\tau|) \cos(\omega_0 \tau), \, -\infty < \tau < \infty.$$

(a) The fact that this is a valid autocorrelation function imposes certain restrictions on c_1 and c_2. What is the range of possible values for c_1? What is the range of possible values for c_2?

(b) What is the expected value of the instantaneous power in $X(t)$?

(c) Consider the random process $Y(t)$ given by $Y(t) = 3X(4t)$, $-\infty < t < \infty$. Is $Y(t)$ wide-sense stationary? Justify your answer.

(d) Find the autocorrelation function for $Y(t)$.

2.21 Consider the system shown in Figure 2–3 of Example 2–2. Suppose that, for the duration of the experiment, the temperature is not held constant, but is instead increased linearly from some initial temperature. The initial temperature at time $t = 0$ is b, so the temperature at time t is given by $T = ct + b$ for some positive constant c.

(a) What is the mean of the random process $Y(t)$ shown in the figure?

(b) Give an expression for the variance of $Y(t)$.

(c) Give an expression for the one-dimensional density function for the random process $Y(t)$.

(d) Does this random process have a constant mean? Is the process stationary?

CHAPTER 3

Linear Filtering of Random Processes

3.0 THE NEED FOR FILTERING OF RANDOM PROCESSES

In electronic systems, it is often necessary to filter random processes in order to improve certain features or remove undesirable characteristics. For example, a particular random process may have "spikes" that will produce unwanted transients in an electronic system. In this situation, filtering of the random process will make it smoother in some sense, such as by eliminating the spikes in the waveform or by decreasing their amplitudes.

If the random process consists of a signal and noise, it is usually necessary to filter the random process to reduce the effects of the noise while minimizing the distortion of the signal. In this case, the filtering can be employed to enhance the signal-to-noise ratio, or it can be used to produce a better quality output according to some other criteria. The need for filtering a random process that consists of a signal and noise arises in every radar, navigation, and communication system.

There are other applications in which the engineer may not have complete control of the filtering operation. An example occurs in the transmission of a signal through the atmosphere, over an optical fiber, or in a waveguide. The propagation medium acts as a filter in these situations, and it is necessary to analyze the effects of the filtering induced by the medium in order to design an efficient receiver. Analysis is also required in order to predict the performance of a system that must operate within such a dispersive medium.

In this chapter, we present the fundamental analytical methods needed to design linear filters and systems to accomplish the various objectives just described. These methods are employed by engineers to determine optimum filters for specific applications, to conduct tradeoff studies and design suboptimum filters, to analyze the performance of existing and proposed electronic systems, and determine the effects of various types of noise and other interference.

The most common performance criteria are based on second-order quantities, such as the mean-squared error, noise power, and signal-to-noise ratio. These are all related to the correlation functions or spectral densities of the noise processes involved. As a result, the relevant analysis of linear filters deals with the relationships between the input and output autocorrelation functions or spectral densities. Accordingly, the chapter is concerned primarily with the determination of the autocorrelation function of the output random process in terms of the autocorrelation function of the input random process and the response function of the linear system. Chapter 4 deals with the spectral densities of the input and output random processes.

3.1   DISCRETE- AND CONTINUOUS-TIME LINEAR SYSTEMS

We begin with a brief review of linear filtering of deterministic signals. Recall that if $x(t)$ is a deterministic continuous-time signal and $h(t)$ is the impulse response of a time-invariant, continuous-time linear system, then the output of the system when x is the input is given by

$$y(t) = \int_{-\infty}^{\infty} h(t - \tau)x(\tau)\, d\tau = \int_{-\infty}^{\infty} h(\tau)x(t - \tau)\, d\tau. \tag{3.1}$$

That is, y is the convolution of x with h, denoted by $y = x * h$. According to (3.1),

$$x * h = h * x.$$

If $x(k)$ is a deterministic discrete-time signal and $h(k)$ is the pulse response of a time-invariant discrete-time linear system, then

$$y(k) = \sum_{n=-\infty}^{\infty} h(k - n)x(n) = \sum_{n=-\infty}^{\infty} h(n)x(k - n) \tag{3.2}$$

is the output of the system when x is the input. Again, y is the convolution of x with h and $x * h = h * x$. In writing discrete-time signals, we always normalize the time scale so that $x(k)$ physically represents the signal at time kt_0. That is, time is measured as multiples of t_0 time units.

Example 3–1   A Simple Linear Discrete-Time Filter

Suppose that the pulse response of a discrete-time system is given by

$$h(k) = \begin{cases} 1, & k = 0 \text{ or } 1 \\ 0, & \text{otherwise} \end{cases}.$$

Then, for any input signal $x(k)$, (3.2) implies that

$$y(k) = \sum_{n=0}^{1} h(n)x(k - n)$$
$$= x(k) + x(k - 1).$$

Notice that this is the same linear system as in Example 2–4; however, the input is a random process in that example, whereas it is a deterministic signal here. This linear filter is illustrated in Figure 3–1. ∎

The input to a discrete-time linear system is often a sampled version $x_s(k)$ of a continuous-time signal $x(t)$. That is, $x_s(k) = x(kt_0)$ for each integer k, where t_0 is the

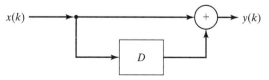

FIGURE 3–1   A simple discrete-time, linear, time-invariant filter.

sampling interval (i.e., the time between samples). Similarly, a continuous-time signal $x_c(t)$ may be constructed from a discrete-time signal $x(k)$. For example, $x_c(t) = x(k)$ for $kt_0 \leq t < (k + 1)t_0$; that is, $x_c(t)$ is a sequence of rectangular pulses, each of duration t_0, and the amplitude of the kth pulse is $x(k)$. The continuous-time signal $x_c(t)$ can be written in terms of the discrete-time signal $x(k)$ as

$$x_c(t) = \sum_{k=-\infty}^{\infty} x(k)p_{t_0}(t - kt_0),$$

where $p_\lambda(t)$ denotes the rectangular pulse of unit amplitude and duration λ that begins at time $t = 0$: $p_\lambda(u) = 1$ for $0 \leq u < \lambda$ and $p_\lambda(u) = 0$ otherwise.

With the preceding relationship between continuous-time signals and discrete-time signals, continuous-time signals can be inputs to discrete-time systems, and discrete-time signals can be inputs to continuous-time systems. Of course, the signals must be converted to the appropriate form first (e.g., the continuous-time signal is initially sampled, and then it is processed by a discrete-time system).

In general, a linear system need not be time invariant. That is, the response at time t due to an input impulse occurring at time λ may depend on both t and λ, rather than on the difference $t - \lambda$ only, as is true for time-invariant linear systems. If the continuous-time signal x is the input to a linear system for which the response at time t to an impulse occurring at time λ is $h(t, \lambda)$, then

$$y(t) = \int_{-\infty}^{\infty} h(t, \tau)x(\tau) \, d\tau \tag{3.3}$$

is the output. As a check, let the input be an impulse occurring at time λ; the corresponding output is, of course, the impulse response of the system. Accordingly, we replace $x(\tau)$ with $\delta(\tau - \lambda)$ in the integrand of (3.3), and this gives $y(t) = h(t, \lambda)$ as the output of the filter, as it should be. If the linear system is time invariant, then $h(t, \tau)$ depends on the time difference $t - \tau$ only, and the impulse response $h(t, \tau)$ is denoted by $h(t - \tau)$. Hence, (3.3) reduces to (3.1) if the linear filter is time invariant.

Example 3–2   The Integrate-and-Dump Filter

The integrate-and-dump filter is a continuous-time, time-varying, linear filter. It is an active filter that can be implemented as an integrator together with a pair of switches. One switch, which is internal to the circuit, closes at time $t = 0$ to eliminate any energy stored from previous intervals, and the other switch, which is at the output of the integrator, closes at time $t = T$ to sample the output. Mathematically, the integrate-and-dump filter is just a finite-time-duration integration process: The output is the integral of the input over the interval $[0, T]$, as illustrated in Figure 3–2.

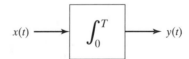

FIGURE 3–2   Integrate-and-dump filter.

The input–output relation for the integrate-and-dump filter is as follows: If x is the input signal, the output is given by $y(t) = 0$ for $t \neq T$ and

$$y(T) = \int_0^T x(u) \, du.$$

What is the impulse response for this filter? By definition, the impulse response $h(t, \tau)$ is the response at time t to an impulse at time τ, so if $x(u) = \delta(u - \tau)$ is the input, the corresponding output is $h(t, \tau)$. First, observe from the foregoing description that $h(t, \tau) = 0$ for $t \neq T$. For $t = T$, the output $h(T, \tau)$ is given by the preceding integral expression with $x(u)$ replaced by $\delta(u - \tau)$; that is,

$$h(T, \tau) = \int_0^T \delta(u - \tau) \, du.$$

But the value of this integral is zero unless the impulse occurs within the limits of integration. So $h(T, \tau) = 0$ for $\tau < 0$ and for $\tau > T$. For τ within the limits of integration (i.e., $0 < \tau < T$), the value of the integral is unity, so $h(T, \tau) = 1$ for $0 < \tau < T$. For $\tau = 0$ or $\tau = T$, the value depends on the convention being followed regarding the value of an integral of a delta function located at one of the endpoints of the interval of integration. Common conventions are to define the value to be zero at one or both endpoints, unity at one or both endpoints, or $\frac{1}{2}$ at each endpoint. For our purposes, it makes no difference how the value of the integral is defined, and it is sufficient to leave it undefined for now. Summarizing these results, we have shown that the impulse response of the integrate-and-dump filter is

$$h(t, \tau) = \begin{cases} 1, & t = T \text{ and } 0 < \tau < T, \\ 0, & t \neq T \text{ or } \tau < 0 \text{ or } \tau > T. \end{cases}$$
∎

For a general *discrete-time* system, the output $h(k, i)$ at time k for a unit-amplitude input pulse at time i depends on k and i. Thus, the more general version of (3.2) is

$$y(k) = \sum_{n=-\infty}^{\infty} h(k, n) x(n). \tag{3.4}$$

Again as a check, suppose x is a unit pulse occurring at time i. That is, $x(i) = 1$ and $x(n) = 0$ for $n \neq i$. Then (3.4) implies that $y(k) = h(k, i)$, as required.

A linear system is *stable* if, for any input that is bounded, the output of the system is also bounded. This is actually just one form of stability, known as *bounded-input, bounded-output stability*, but it is the only form considered in this book. Notice that the filtering operations characterized by (3.1) and (3.3) represent stable continuous-time linear systems if, for any finite positive number M_x, $|x(\tau)| \leq M_x$ for all τ implies that there is a number $M_y < \infty$ such that $|y(t)| \leq M_y$ for all t. Similarly, (3.2) and (3.4) represent stable discrete-time linear systems if, for any $M_x < \infty$, $|x(n)| \leq M_x$ for all n implies that there is a number $M_y < \infty$ such that $|y(k)| \leq M_y$ for all k.

Notice that if $|x(\tau)| \leq M_x$ in (3.3), then

$$|y(t)| \leq \int_{-\infty}^{\infty} |h(t, \tau)||x(\tau)| \, d\tau$$

$$\leq M_x \int_{-\infty}^{\infty} |h(t, \tau)| \, d\tau.$$

Thus, a sufficient condition for stability is the existence of a number $M_h < \infty$ such that

$$\int_{-\infty}^{\infty} |h(t, \tau)| \, d\tau \leq M_h, \quad \text{for all } t. \tag{3.5}$$

If (3.5) holds, we can simply let $M_y = M_x M_h$. On the other hand, if the system is stable, then it must be that

$$\int_{-\infty}^{\infty} |h(t, \tau)| \, d\tau < \infty, \quad \text{for all } t. \tag{3.6}$$

Otherwise, for a fixed t, we can simply let $x(\tau) = +1$ for values of τ for which $h(t, \tau) \geq 0$ and $x(\tau) = -1$ for those τ for which $h(t, \tau) < 0$. For this bounded input, the output at time t will be infinite.

Notice that if a system is time invariant, then (3.5) and (3.6) are equivalent, and they reduce to

$$\int_{-\infty}^{\infty} |h(\lambda)| \, d\lambda < \infty. \tag{3.7}$$

That is, (3.7) is both necessary and sufficient for the stability of a time-invariant, continuous-time, linear system.

The discrete-time counterpart to (3.5) is the existence of a number $M_h < \infty$ such that

$$\sum_{n=-\infty}^{\infty} |h(k, n)| \leq M_h, \quad \text{for all } k, \tag{3.8}$$

which is a sufficient condition for the stability of a discrete-time linear system. A necessary condition is

$$\sum_{n=-\infty}^{\infty} |h(k, n)| < \infty, \quad \text{for all } k. \tag{3.9}$$

These two conditions reduce to

$$\sum_{i=-\infty}^{\infty} |h(i)| < \infty \tag{3.10}$$

if the discrete-time linear system is time invariant. Thus, (3.10), which is analogous to (3.7), is a necessary and sufficient condition for the stability of a time-invariant, discrete-time, linear system.

An important observation to be made concerning (3.1)–(3.4) is that the mathematical operation that appears on the right-hand side of each of these equations is a *limit of finite linear combinations* of values of the input signal. For instance, (3.4) is just the statement that, for each k,

$$y(k) = \lim_{N \to \infty} y_N(k), \tag{3.11}$$

where

$$y_N(k) = \sum_{n=-N}^{N} h(k, n) x(n). \tag{3.12}$$

Similarly, (3.3) actually means that, for each t,

$$y(t) = \lim_{N \to \infty} y_N(t), \tag{3.13}$$

where

$$y_N(t) = \int_{-N}^{N} h(t, \tau) x(\tau) \, d\tau, \tag{3.14}$$

which implies

$$y_N(t) = \lim_{m \to \infty} m^{-1} \sum_{k=-Nm}^{Nm} h(t, km^{-1}) x(km^{-1}). \tag{3.15}$$

Notice that (3.15) is just the definition of the Riemann integral that appears in (3.14); it may look more familiar if m^{-1} is replaced by $\Delta \tau$. The limits in (3.11), (3.13), and (3.15) always exist for stable systems with bounded inputs.

The reason for discussing the mathematical meaning of (3.1)–(3.4) is to emphasize that these linear operations are defined as *limits of finite linear combinations* of values of the input signal. That is, both (3.3) and (3.4) involve limits of sequences of quantities of the form

$$\sum_{k=1}^{K} c_k x(t_k),$$

where the coefficient c_k is a real number and each t_k is an integer (for discrete time) or a real number (for continuous time). The coefficients c_k depend on the impulse responses of the system, but not on the input signal; the parameters t_k represent the times at which the signal is sampled in order to form the finite linear combination.

3.2   LINEAR OPERATIONS ON RANDOM PROCESSES

Suppose now that the input to the discrete-time linear system described by (3.4) is a discrete-time random process $X(k)$, $k \in \mathbb{Z}$. The output random process $Y(k)$, $k \in \mathbb{Z}$, can be described by an infinite series, which we denote by

$$Y(k) = \sum_{n=-\infty}^{\infty} h(k, n) X(n). \tag{3.16}$$

However, simply writing (3.16) does not define $Y(k)$, because the right-hand side is an infinite series of random variables, which is a limit of finite sums of random variables. Does such a limit exist? If so, in what sense?

We would like to define the infinite series in (3.16) in a manner analogous to the way we defined it in (3.11) and (3.12), except that the approximating finite sums

$$Y_N(k) = \sum_{n=-N}^{N} h(k, n) X(n) \tag{3.17}$$

are now *random variables*. It turns out that under certain conditions, the limit does exist in the sense that there exists a second-order random process $Y(k)$ for which

$$\lim_{N \to \infty} E\{[Y(k) - Y_N(k)]^2\} = 0. \tag{3.18}$$

According to (3.18), the *mean* of the *square* of the difference between $Y(k)$ and $Y_N(k)$ must converge to zero as $N \to \infty$. Hence, the infinite series in (3.16) is defined as a mean-square infinite series. If (3.18) holds, we say that, for each k, $Y_N(k)$ converges to $Y(k)$ in a *mean-square sense*.

For deterministic signals, we require only that

$$\lim_{N \to \infty} |y(k) - y_N(k)| = 0, \tag{3.19}$$

which follows from (3.11). Therefore, it is reasonable to ask why we employ (3.18) to define convergence, rather than

$$\lim_{N \to \infty} E\{|Y(k) - Y_N(k)|\} = 0. \tag{3.20}$$

In fact, it turns out that (3.18) implies (3.20), but the converse is not true. More importantly, (3.20) is not sufficient to handle the type of problems that we treat in subsequent sections. We shall be concerned primarily with second-order properties (e.g., properties derived from correlation functions), and the study of second-order properties of the output of a linear system with a random-process input requires that the linear operation be defined in a *mean-square* sense.

The mean-square limit of the sequence $Y_N(k)$ need not exist for all input random processes and discrete-time linear filters. However, the conditions sufficient for the existence of a second-order random process $Y(k)$ satisfying (3.16) are met by most random processes and filters encountered in practical applications. These conditions are given in terms of the function $S(\cdot)$ defined by

$$S(k) = \sum_{i=-\infty}^{\infty} \sum_{n=-\infty}^{\infty} |h(k,n)h(k,i)R_X(n,i)|, \quad \text{for each } k. \tag{3.21a}$$

It can be shown that the mean-square limit $Y(k)$ exists for each k if and only if the process $X(k)$ has an autocorrelation function that satisfies

$$S(k) < \infty, \quad \text{for each } k. \tag{3.21b}$$

If the system is stable and if there exists a real number r for which

$$R_X(n,n) \leq r, \quad \text{for each } n, \tag{3.22}$$

then (3.21) is satisfied.* To show that this is true, we first use the bound of (2.17), along with (3.22), to conclude that

$$|R_X(n,i)| \leq \sqrt{R_X(n,n)R_X(i,i)} \leq r. \tag{3.23}$$

It follows from (3.21a) and (3.23) that

$$S(k) \leq r \left\{ \sum_{i=-\infty}^{\infty} |h(k,i)| \right\} \left\{ \sum_{n=-\infty}^{\infty} |h(k,n)| \right\},$$

which must be finite for all k because of (3.9). Hence, for an input random process $X(k)$ with bounded second moment, the limit required by (3.16) always exists in a mean-square sense if the discrete-time system is stable. Notice that (3.22) is always satisfied for $r = R_X(0)$ if $X(k)$ is a second-order, wide-sense stationary random process.

---

*The statement that r is a real number is meant to imply that r is finite. The numbers $+\infty$ and $-\infty$ are usually referred to as *extended* real numbers.

If the input to a continuous-time system is a continuous-time random process $X(t)$, then the output is a continuous-time process described by the stochastic integral

$$Y(t) = \int_{-\infty}^{\infty} h(t, \tau) X(\tau) \, d\tau. \tag{3.24}$$

This integral is defined by taking limits of finite linear combinations, just as in (3.13)–(3.15), except that the limits are defined in a mean-square sense analogous to the way they are defined in (3.16)–(3.18). A sufficient condition for $Y(t)$ in (3.24) to be defined as a mean-square limit is

$$\int_{-\infty}^{\infty} \int_{-\infty}^{\infty} |h(t, t_1) h(t, t_2) R_X(t_1, t_2)| \, dt_1 \, dt_2 < \infty, \quad \text{for each } t, \tag{3.25}$$

by which we mean that this Riemann integral must exist and be finite.

For random processes encountered in practice, the stability of the system and the condition that there is a real number r for which

$$R_X(t, t) \le r, \quad \text{for each } t, \tag{3.26}$$

are enough to ensure that the Riemann integral in (3.25) is defined. Given that the integral is defined, these two conditions also guarantee that the integral is finite. If $X(t)$ is a second-order wide-sense stationary process, then (3.26) is always satisfied for $r = R_X(0)$.

In general, by a *linear operation on a random process*, we mean either a finite linear combination of the random variables that make up the process or a mean-square limit of a sequence of such linear combinations. Most of the linear operations that are of interest to us are of the form of (3.16), (3.17), or (3.24). There are a few important operations not easily described by one of these three forms, however. One example is the differentiation of a continuous-time random process—that is,

$$Y(t) = X'(t) = \frac{dX(t)}{dt},$$

which is defined as a mean-square limit of difference quotients. That is, $Y(t) = X'(t)$ satisfies

$$\lim_{n \to \infty} E\left[Y(t) - \frac{X(t + s_n) - X(t)}{s_n} \right]^2 = 0$$

for any sequence $\{s_n\}$ converging to zero. Clearly, if

$$Y_n(t) = \frac{X(t + s_n) - X(t)}{s_n},$$

then $Y(t)$ is the mean-square limit of $Y_n(t)$, and $Y_n(t)$ is a finite linear combination of the random variables $\{X(t): -\infty < t < \infty\}$. Thus, the differentiation of $X(t)$ is a linear operation easily described as a mean-square limit of finite linear combinations of the random variables in the collection $\{X(t): -\infty < t < \infty\}$, but it is not conveniently expressible as (3.16), (3.17), or (3.24).

3.3   TIME-DOMAIN ANALYSIS OF SECOND-ORDER RANDOM PROCESSES IN LINEAR SYSTEMS

In most engineering investigations, such as the study of noise in electronic systems, the information about a random process that is most readily obtained and, for many problems, most useful is that which is provided by the mean and autocorrelation functions of the random process. Knowledge of these functions is sufficient to permit evaluation of the *noise power* at some point in a circuit or system or the *signal-to-noise ratio* in a radar, navigation, or communication receiver. These functions are also required in order to obtain certain types of *predictions* of a future value of a random process based on observations of past values or *estimates* of one random process based on observations of another random process.

As discussed at the beginning of this chapter, linear filtering of the random process is often required in order to decrease the power in the random process, improve the signal-to-noise ratio, or accomplish some other desirable objective. To design a linear filter for this purpose, it is necessary to be able to determine the characteristics of the output of the filter as a function of the characteristics of the input and the filter.

In this section, the focus is on the *time-domain* characteristics of the input and output random process and the linear filter. Consequently, we will be working with the mean and autocorrelation functions for the random processes and the appropriate response functions for the linear filters (i.e., impulse response for continuous-time filters or pulse response for discrete-time filters). Specifically, we will show that if the random process $X(t)$ is the input to a linear filter with known response function and $Y(t)$ is the corresponding output, the mean and autocorrelation functions for the output process can be determined, even if only the mean and autocorrelation functions for $X(t)$ are known. In particular, the output mean function can be determined from the input mean function, and the output autocorrelation function can be determined from the input autocorrelation function. Moreover, for the important special case in which $X(t)$ is a Gaussian random process, we can determine a great deal more, as will be discussed in Section 3.4.

We begin by considering the linear filtering of a *continuous-time* random process $X(t)$, $-\infty < t < \infty$. Recall that, under suitable conditions, the output is a continuous-time random process

$$Y(t) = \int_{-\infty}^{\infty} h(t, \tau) X(\tau)\, d\tau.$$

Given *only* the mean and autocorrelation functions for the input process $X(t)$, can the mean and autocorrelation functions for the output process $Y(t)$ be determined? Fortunately, the answer is that they can. In this section, expressions are developed that give the mean function for the output process in terms of the mean function for the input process and the impulse response of the linear system. Similarly, expressions are developed that give the autocorrelation function for the output process in terms of the autocorrelation function for the input process and the impulse response of the linear system.

Analogous expressions will be given for discrete-time linear systems in which the input–output relationship is given by

$$Y(k) = \sum_{n=-\infty}^{\infty} h(k, n) X(n).$$

For a discrete-time linear system with a given pulse response, knowledge of the mean function for the input is sufficient to determine the mean function for the output. Furthermore, in order to find the autocorrelation function for the output process, it suffices to know only the input autocorrelation function.

Before we develop the general expressions, it may be helpful to look at a specific example. In the solution of the exercise that follows, expressions for the mean and autocorrelation functions for the output are developed that turn out to be special cases of the general results presented later in this section. The primary simplification in this exercise stems from the fact that only finite sums are needed, rather than the integrals or infinite sums that are required in the analysis of general continuous-time or discrete-time linear systems.

Exercise 3–1. Suppose a discrete-time, time-invariant, linear system has pulse response

$$h(n) = \begin{cases} 1, & n = 1, \\ \frac{1}{2}, & n = 0 \text{ or } n = 2, \\ 0, & \text{otherwise.} \end{cases}$$

Let $Y(k)$ be the output of this system when the input is a second-order random process $X(k)$ for which the mean is $\mu_X(k) = E\{X(k)\}$ and the autocorrelation is $R_X(k, i) = E\{X(k)X(i)\}$. Find the mean and autocorrelation functions for the output.

Solution. Since $h(n) = 0$ except for n in the range $0 \le n \le 2$, the output is given by

$$Y(k) = \sum_{n=0}^{2} h(n)X(k - n) = \left(\tfrac{1}{2}\right)X(k) + X(k - 1) + \left(\tfrac{1}{2}\right)X(k - 2).$$

Thus, the mean of the output is

$$\mu_Y(k) = E\{Y(k)\} = \sum_{n=0}^{2} h(n)E\{X(k - n)\}$$

$$= \sum_{n=0}^{2} h(n)\mu_X(k - n)$$

$$= \left(\tfrac{1}{2}\right)\mu_X(k) + \mu_X(k - 1) + \left(\tfrac{1}{2}\right)\mu_X(k - 2).$$

The autocorrelation function for the output process is

$$R_Y(k, i) = E\{Y(k)Y(i)\}$$

$$= E\left\{ \sum_{n=0}^{2} h(n)X(k - n) \sum_{m=0}^{2} h(m)X(i - m) \right\}$$

$$= E\left\{ \sum_{n=0}^{2} \sum_{m=0}^{2} h(n)h(m)X(k - n)X(i - m) \right\}$$

$$= \sum_{n=0}^{2} \sum_{m=0}^{2} h(n)h(m)E\{X(k - n)X(i - m)\}.$$

Replacing the expectation of the product by the autocorrelation function, we find that

$$R_Y(k, i) = \sum_{n=0}^{2} \sum_{m=0}^{2} h(n)h(m)R_X(k - n, i - m)$$

$$= \left(\tfrac{1}{4}\right)[R_X(k, i) + R_X(k, i - 2) + R_X(k - 2, i) + R_X(k - 2, i - 2)] + \left(\tfrac{1}{2}\right)[R_X(k, i - 1)$$
$$+ R_X(k - 1, i) + R_X(k - 1, i - 2) + R_X(k - 2, i - 1)] + R_X(k - 1, i - 1).$$

If the input process is wide-sense stationary, the expressions for the mean and autocorrelation functions simplify to $\mu_Y(k) = \mu_Y = 2\mu_X$ and

$$
\begin{aligned}
R_Y(k,i) = R_Y(k - i) &= \left(\tfrac{1}{4}\right)[2R_X(k - i) + R_X(k - i + 2) + R_X(k - i - 2)] \\
&\quad + \left(\tfrac{1}{2}\right)[2R_X(k - i + 1) + 2R_X(k - i - 1)] + R_X(k - i) \\
&= \left(\tfrac{3}{2}\right)R_X(k - i) + \left(\tfrac{1}{4}\right)[R_X(k - i + 2) + R_X(k - i - 2)] \\
&\quad + R_X(k - i + 1) + R_X(k - i - 1).
\end{aligned}
\tag{3.27}
$$

Since $\mu_Y(k)$ does not depend on k and $R_Y(k,i)$ depends on the difference $k - i$ only, the output process is also wide-sense stationary.

As a special case of (3.27), obtained by setting $i = k$, we have

$$
E[Y(k)]^2 = R_Y(0) = \left(\tfrac{3}{2}\right)R_X(0) + 2R_X(1) + \left(\tfrac{1}{2}\right)R_X(2),
\tag{3.28}
$$

where we have used the property $R_X(\tau) = R_X(-\tau)$. From (3.28), we see that the variance of the output process can be written as

$$
\begin{aligned}
\sigma_Y^2 = E[Y(k) - \mu_Y]^2 &= \left(\tfrac{3}{2}\right)R_X(0) + 2R_X(1) + \left(\tfrac{1}{2}\right)R_X(2) - 4\mu_X^2 \\
&= \left(\tfrac{3}{2}\right)C_X(0) + 2C_X(1) + \left(\tfrac{1}{2}\right)C_X(2),
\end{aligned}
$$

where we have used the fact that $R_X(\tau) - \mu_X^2 = C_X(\tau)$ in the second step. The result for the variance is equivalent to

$$
\sigma_Y^2 = \left(\tfrac{3}{2}\right)\sigma_X^2 + 2C_X(1) + \left(\tfrac{1}{2}\right)C_X(2).
\tag{3.29}
$$

Notice that although the output mean depends only on the input mean and the output autocorrelation function depends only on the input autocorrelation function, *the output variance does not depend only on the input variance.* As shown in (3.29), certain values of the input autocovariance function must be known in order to compute the output variance. Similarly, (3.28) shows that the second moment of the output depends on certain values of the input autocorrelation function. This fact is one of the reasons that much of our attention has been focused on the analysis of the mean and autocorrelation function (or the mean and autocovariance function) for the input and output random processes. ∎

We now develop the general relationships between input and output mean and autocorrelation functions. The complete derivation is given for continuous-time random processes and continuous-time linear systems. The derivations are analogous for discrete-time linear systems—just replace integrals by infinite sums—so only the final results are given. The results are for general, linear, time-varying systems, from which the results for time-invariant systems are obtained as a special case. It is assumed throughout that (3.21) and (3.25) are satisfied, so that the input–output equations [i.e., (3.16) and (3.24)] are well defined in a mean-square sense.

3.3.1 Mean of the Output of a Linear System

If the input process $X(t)$ to a continuous-time linear system with impulse response $h(t, \tau)$ has mean $\mu_X(t)$, then the output process

$$
Y(t) = \int_{-\infty}^{\infty} h(t, \tau)X(\tau) \, d\tau
$$

has mean

$$\mu_Y(t) = E\{Y(t)\} = E\left\{\int_{-\infty}^{\infty} h(t,\tau)X(\tau)\,d\tau\right\}$$

$$= \int_{-\infty}^{\infty} h(t,\tau)E\{X(\tau)\}\,d\tau$$

$$= \int_{-\infty}^{\infty} h(t,\tau)\mu_X(\tau)\,d\tau. \tag{3.30}$$

A *sufficient* condition for the interchange of expectation and integration to be valid is that the linear system be stable and the input be a second-order random process. In fact, this interchange is valid for virtually all linear systems and random processes that are actually encountered in engineering problems.

Notice in particular that $\mu_Y(t)$ is the output of the system when $\mu_X(t)$ is the (deterministic) input, a result that is certainly not surprising, either from an intuitive point of view or from purely mathematical arguments. The expected value of a sum (or integral) of random variables (or random processes) is simply the sum (or integral) of the expected values of the random variables (or random processes). This fact is an illustration of a situation in which "the expectation operation can be moved inside the integral," a principle that is valid for the vast majority of engineering applications.

More generally, we say that "the order of the expectation operation and a *linear* operation can be interchanged without changing the result." If L denotes such a linear operation, this statement can be written in shorthand form as $E\{L(X(t))\} = L(E\{X(t)\})$. The notation used here is not very precise and is intended only as an aid to intuition. For example, it is more accurate to write $L(\{X(t):t \in T\})$ or $L(\{X_t:t \in T\})$ rather than $L(X(t))$, because, in general, L operates on the entire process $\{X(t):t \in T\}$. The linear operations of interest to us in this section are those described by (3.16), (3.17), or (3.24).

If the linear system is time invariant, (3.30) reduces to

$$\mu_Y(t) = \int_{-\infty}^{\infty} h(t-\tau)\mu_X(\tau)\,d\tau = \int_{-\infty}^{\infty} h(\tau)\mu_X(t-\tau)\,d\tau. \tag{3.31}$$

If, in addition, the mean of the random process $X(t)$ is not a function of t (i.e., $E\{X(t)\} = \mu_X$, a constant), then (3.31) reduces further to

$$\mu_Y(t) = \mu_Y = \mu_X \int_{-\infty}^{\infty} h(\tau)\,d\tau. \tag{3.32}$$

Thus, given an input process $X(t)$ with a finite, constant mean, a sufficient condition for the output of a time-invariant linear system with impulse response $h(t)$ to have a finite, constant mean is

$$\int_{-\infty}^{\infty} |h(t)|\,dt < \infty. \tag{3.33}$$

This is just the condition for stability of the time-invariant linear system.

On the basis of the results of the preceding analysis, we can state some general properties of time-invariant linear systems with random-process inputs. These properties are valid for both discrete- and continuous-time systems and processes:

1. *The output mean can be determined from the input mean alone; higher order moments are not required.*
2. *If the input mean is constant, so is the output mean.*
3. *If the input mean is constant and finite and the system is stable, the output mean is constant and finite.*

These properties do *not* hold for general nonlinear systems.

If the mean of the input process $X(t)$ is not constant and the linear system is not time invariant, the condition for finiteness of the output mean $\mu_Y(t)$ is

$$\int_{-\infty}^{\infty} |h(t,\tau)| |\mu_X(\tau)| \, d\tau < \infty. \tag{3.34}$$

Condition (3.34) reduces to (3.33) if the system is time invariant, in which case $h(t,\tau) = h(t-\tau)$ for all t and τ, and if $\mu_X(t) = \mu_X$, a finite constant. Condition (3.34) is also sufficient to ensure the validity of the interchange in the order of expectation and integration that is employed in (3.30).

Exercise 3–2. A time-invariant linear system has impulse response

$$h(t) = \begin{cases} 0, & t < 0, \\ 1, & 0 \le t < T_0, \\ 0, & t \ge T_0. \end{cases}$$

Suppose the input to this system is the random process defined by $X(t) = Y_1 + tY_2$ for $-\infty < t < \infty$, and Y_1 and Y_2 are random variables with finite means ν_1 and ν_2, respectively. Find the mean of the output process $Y(t)$.

Solution. The mean of $Y(t)$ is given by

$$\mu_Y(t) = \int_0^{T_0} \mu_X(t - \tau) \, d\tau.$$

We use the fact that $\mu_X(t) = \nu_1 + \nu_2 t$ to deduce

$$\mu_Y(t) = \nu_1 T_0 + \int_0^{T_0} (t - \tau)\nu_2 \, d\tau.$$

Evaluating the integral, we see that $\mu_Y(t) = \left[\nu_1 T_0 - \left(\frac{1}{2}\right)\nu_2 T_0^2\right] + (\nu_2 T_0)t$. Notice that the output mean is of the form $a + tb$ where $a = \nu_1 T_0 - \left(\frac{1}{2}\right)\nu_2 T_0^2$ and $b = \nu_2 T_0$. The special case $\nu_2 = 0$ gives a constant input mean $\mu_X(t) = \nu_1$ and a constant output mean $\mu_Y(t) = \nu_1 T_0$. Notice also that both (3.33) and (3.34) are satisfied in this example (even if $\nu_2 \ne 0$). ∎

The key results for *continuous-time* linear systems and processes are (3.30), (3.31), and (3.32). For *discrete-time* linear systems, the analogous expressions are as follows: In general,

$$\mu_Y(k) = \sum_{i=-\infty}^{\infty} h(k,i)\mu_X(i). \tag{3.35}$$

For time-invariant systems (3.35) reduces to

$$\mu_Y(k) = \sum_{i=-\infty}^{\infty} h(k-i)\mu_X(i) = \sum_{i=-\infty}^{\infty} h(i)\mu_X(k-i), \tag{3.36}$$

which in turn yields

$$\mu_Y(k) = \mu_Y = \mu_X \sum_{i=-\infty}^{\infty} h(i) \tag{3.37}$$

if $X(k)$ has constant mean μ_X. The condition for the finiteness of $\mu_Y(k)$ in (3.35) is

$$\sum_{i=-\infty}^{\infty} |h(k,i)||\mu_X(i)| < \infty. \tag{3.38}$$

The mean μ_Y in (3.37) is finite if μ_X is finite and the system is stable.

Example 3–3   A Time-Varying Discrete-Time Linear Filter

Consider a discrete-time linear system with pulse response

$$h(k,i) = \begin{cases} k2^{-(k-i)}, & k \geq i, \\ 0, & k < i. \end{cases}$$

The input to this system is a wide-sense stationary process $X(k)$ with mean $\mu_X > 0$. The mean of the output process $Y(k)$ is given by (3.35), with $\mu_X(i) = \mu_X$. That is,

$$\mu_Y(k) = \mu_X \sum_{i=-\infty}^{\infty} h(k,i) = \mu_X \sum_{i=-\infty}^{k} k2^{-(k-i)},$$

so the output mean is given by

$$\mu_Y(k) = k\mu_X \sum_{i=-\infty}^{k} 2^{-(k-i)}.$$

Making the substitution $n = k - i$, we find that

$$\mu_Y(k) = k\mu_X \sum_{n=0}^{\infty} 2^{-n} = 2k\mu_X.$$

Clearly, the output mean $\mu_Y(k)$ is not constant, so the process $Y(k)$ is not wide-sense stationary. In fact, the output mean is not even bounded ($\mu_Y(k) \to \infty$ as $k \to \infty$). This is because the given pulse response does not satisfy (3.8), although it does satisfy (3.9). The system is not stable, as can be seen by considering the bounded input $x(i) = \mu_X$ for all i. The corresponding output is $y(k) = \mu_Y(k) = 2k\mu_X$, which increases without bound as $k \to \infty$ if $\mu_X > 0$. However, the output mean is finite for each k, which is always true if (3.38) is satisfied. (Note that (3.38) is indeed satisfied in this example.) ∎

Exercise 3–3. Suppose a continuous-time linear system has impulse response

$$h(t,\tau) = \begin{cases} \exp\{-\alpha(t-\tau)\}, & t \geq \tau, \\ 0, & t < \tau. \end{cases}$$

If the input process $X(t)$ has constant mean μ_X, find the mean of the output process $Y(t)$. Assume that $\alpha > 0$.

Solution. Since the system is time invariant and the input mean is constant, we employ (3.32) and conclude that

$$\mu_Y = \mu_X \int_0^{\infty} e^{-\alpha\tau}\, d\tau = \alpha^{-1}\mu_X. \quad ∎$$

Exercise 3–4. A discrete-time linear system is described as follows: For any input $x(k)$, the output is the signal

$$y(k) = \left(\tfrac{1}{2}\right)y(k-1) + x(k). \tag{3.39}$$

Suppose the input to this system is a random process $X(k)$ with constant mean μ_X. What is the mean of the output process $Y(k)$?

Solution. We give two methods.

(a) The pulse response of the system described by (3.39) is found by letting $x(i) = 1$ and $x(n) = 0$ for all $n \neq i$. For this input, $y(k) = h(k,i)$. From (3.39), we see that $y(k) = 0$ for $k < i$, $y(i) = x(i) = 1$, and $y(k) = \left(\tfrac{1}{2}\right)y(k-1)$ for $k > i$. Thus, $y(i) = 1, y(i+1) = \tfrac{1}{2}, y(i+2) = \tfrac{1}{4}, \ldots, y(i+m) = 2^{-m}$. The pulse response is therefore

$$h(k,i) = \begin{cases} 2^{-(k-i)}, & k \geq i, \\ 0, & k < i. \end{cases}$$

Hence, the system is time invariant, so (3.37) can be applied and we conclude that the output has constant mean

$$\mu_Y = \mu_X \sum_{n=0}^{\infty} 2^{-n} = 2\mu_X.$$

(b) For the second method, we notice first that the linear system described by (3.39) is time invariant (since, if $w(k)$ is the output signal for input $v(k)$, then $w(k-m)$ is the output signal whenever $v(k-m)$ is the input). This implies that the output

$$Y(k) = \left(\tfrac{1}{2}\right)Y(k-1) + X(k)$$

has constant mean μ_Y. Taking expectations, we find that $\mu_Y = \left(\tfrac{1}{2}\right)\mu_Y + \mu_X$, so $\mu_Y = 2\mu_X$.   ∎

3.3.2   Autocorrelation Function for the Output of a Linear System

If $Y(t)$ is the output of a continuous-time linear system with impulse response $h(t, \tau)$ and input $X(t)$, then the crosscorrelation function for the processes $Y(t)$ and $X(t)$ is given by

$$\begin{aligned}
R_{Y,X}(t,s) &= E\{Y(t)X(s)\} \\
&= E\left\{\int_{-\infty}^{\infty} h(t,\tau)X(\tau)\, d\tau\, X(s)\right\} \\
&= E\left\{\int_{-\infty}^{\infty} h(t,\tau)X(\tau)X(s)\, d\tau\right\}.
\end{aligned}$$

Interchanging the order of expectation and integration, we see that

$$R_{Y,X}(t,s) = \int_{-\infty}^{\infty} h(t,\tau)R_X(\tau,s)\, d\tau. \tag{3.40}$$

The autocorrelation function for the output process $Y(t)$ is

$$R_Y(t, s) = E\{Y(t)Y(s)\}$$

$$= E\left\{Y(t) \int_{-\infty}^{\infty} h(s, \lambda)X(\lambda)\, d\lambda\right\}$$

$$= \int_{-\infty}^{\infty} h(s, \lambda)E\{Y(t)X(\lambda)\}\, d\lambda$$

$$= \int_{-\infty}^{\infty} h(s, \lambda)R_{Y,X}(t, \lambda)\, d\lambda. \tag{3.41}$$

Combining (3.40) and (3.41), we find that

$$R_Y(t, s) = \int_{-\infty}^{\infty} \int_{-\infty}^{\infty} h(s, \lambda)h(t, \tau)R_X(\tau, \lambda)\, d\tau\, d\lambda. \tag{3.42}$$

Note in particular that for the determination of the autocorrelation function for the output process, knowledge of the system impulse response and the autocorrelation function for the input process is sufficient. The same is true for autocovariance functions, since the foregoing derivation can easily be modified to establish the relationship

$$C_Y(t, s) = \int_{-\infty}^{\infty} \int_{-\infty}^{\infty} h(s, \lambda)h(t, \tau)C_X(\tau, \lambda)\, d\tau\, d\lambda. \tag{3.43}$$

Alternatively, (3.43) can be obtained from (3.42), (3.30) and (2.9).

If the input process is wide-sense stationary, (3.40) can be written as

$$R_{Y,X}(t, s) = \int_{-\infty}^{\infty} h(t, \tau)R_X(\tau - s)\, d\tau$$

$$= \int_{-\infty}^{\infty} h(t, \tau)R_X(s - \tau)\, d\tau$$

$$= \int_{-\infty}^{\infty} h(t, s - \alpha)R_X(\alpha)\, d\alpha,$$

which is a convolution integral: For each t, the function $R_{Y,X}(t, \cdot)$ is the convolution of the function $R_X(\cdot)$ with the function $h(t, \cdot)$.

Another important special case results if the system is time invariant. In this case, (3.40) can be written as

$$R_{Y,X}(t, s) = \int_{-\infty}^{\infty} h(t - \tau)R_X(\tau, s)\, d\tau$$

$$= \int_{-\infty}^{\infty} h(\alpha)R_X(t - \alpha, s)\, d\alpha, \tag{3.44}$$

and (3.41) reduces to

$$R_Y(t, s) = \int_{-\infty}^{\infty} h(s - \lambda) R_{Y,X}(t, \lambda) \, d\lambda$$

$$= \int_{-\infty}^{\infty} h(\beta) R_{Y,X}(t, s - \beta) \, d\beta. \tag{3.45}$$

Thus, for time-invariant systems, (3.42) becomes

$$R_Y(t, s) = \int_{-\infty}^{\infty} \int_{-\infty}^{\infty} h(s - \lambda) h(t - \tau) R_X(\tau, \lambda) \, d\tau \, d\lambda$$

$$= \int_{-\infty}^{\infty} \int_{-\infty}^{\infty} h(\alpha) h(\beta) R_X(t - \alpha, s - \beta) \, d\alpha \, d\beta.$$

Finally, if $X(t)$ is wide-sense stationary and the system is time invariant, then

$$R_{Y,X}(t, s) = \int_{-\infty}^{\infty} h(t - \tau) R_X(\tau - s) \, d\tau$$

$$= \int_{-\infty}^{\infty} h(t - s - \alpha) R_X(\alpha) \, d\alpha. \tag{3.46}$$

Notice that $R_{Y,X}(t, s)$ is a function of t and s via the difference $t - s$ only; that is, $R_{Y,X}(t, s) = R_{Y,X}(t - s, 0) = R_{Y,X}(t - s)$ for all t and s. Also, we see that in this case (3.45) and (3.46) imply that

$$R_Y(t, s) = \int_{-\infty}^{\infty} h(\beta) R_{Y,X}(t, s - \beta) \, d\beta$$

$$= \int_{-\infty}^{\infty} \int_{-\infty}^{\infty} h(\alpha) h(\beta) R_X(t - \alpha, s - \beta) \, d\alpha \, d\beta$$

$$= \int_{-\infty}^{\infty} \int_{-\infty}^{\infty} h(\alpha) h(\beta) R_X(t - s + \beta - \alpha) \, d\alpha \, d\beta. \tag{3.47}$$

so that $R_Y(t, s) = R_Y(t - s, 0) = R_Y(t - s)$ for all t and s. Notice that (3.32), (3.46), and (3.47) establish the following important fact:

> *If the input to a time-invariant linear system is a wide-sense stationary random process, the output is also a wide-sense stationary random process; moreover, the input and output processes are jointly wide-sense stationary.*

For this situation, we can write (3.46) as

$$R_{Y,X}(\tau) = \int_{-\infty}^{\infty} h(\tau - \alpha) R_X(\alpha) \, d\alpha$$

$$= \int_{-\infty}^{\infty} h(\alpha) R_X(\tau - \alpha) \, d\alpha \tag{3.48}$$

and (3.47) as

$$R_Y(\tau) = \int_{-\infty}^{\infty} h(\beta) \left\{ \int_{-\infty}^{\infty} h(\alpha) R_X[(\tau + \beta) - \alpha] \, d\alpha \right\} \, d\beta. \tag{3.49}$$

Notice that (3.48) is just the convolution of h with R_X; that is, $R_{Y,X} = h * R_X$.

The outer integral in (3.49) is not quite a convolution integral; it is given by

$$R_Y(\tau) = \int_{-\infty}^{\infty} h(\beta) R_{Y,X}(\tau + \beta) \, d\beta, \tag{3.50}$$

where we have replaced τ by $\tau + \beta$ in (3.48) and substituted for the term in brackets in (3.49). However, if we let $\widetilde{h}$ be the *time-reverse* impulse response (i. e., $\widetilde{h}(t) = h(-t)$ for all t), then (3.50) becomes

$$R_Y(\tau) = \int_{-\infty}^{\infty} \widetilde{h}(-\beta) R_{Y,X}(\tau + \beta) \, d\beta$$

$$= \int_{-\infty}^{\infty} \widetilde{h}(\lambda) R_{Y,X}(\tau - \lambda) \, d\lambda.$$

That is,

$$R_Y = \widetilde{h} * R_{Y,X} = \widetilde{h} * (h * R_X).$$

Since the order in which the convolutions are performed does not alter the answer, we write

$$R_Y = \widetilde{h} * (h * R_X) = (\widetilde{h} * h) * R_X. \tag{3.51}$$

So, in fact, we can write $R_Y = \widetilde{h} * h * R_X$ with no fear of ambiguity. The preceding considerations suggest one approach to finding the output autocorrelation: First convolve the impulse response with its time reverse, and then convolve the result with the input autocorrelation function. Notice that the definition of $f = \widetilde{h} * h$ is the integral

$$f(u) = \int_{-\infty}^{\infty} \widetilde{h}(u - v) h(v) \, dv,$$

but this is equivalent to

$$f(u) = \int_{-\infty}^{\infty} h(v - u) h(v) \, dv, \tag{3.52}$$

because $\widetilde{h}(t) = h(-t)$ for any t. Replacing the variables u and v by τ and t, respectively, and reversing the order of the terms in the integrand, we see that (3.52) becomes

$$f(\tau) = \int_{-\infty}^{\infty} h(t) h(t - \tau) \, dt, \tag{3.53}$$

which is just the integral of the product of the impulse response with a delayed version of the impulse response. The expression in (3.53) is much easier to work with than the general convolution integral, because (3.53) requires only the integration of the product of a function and a time-shifted version of itself; there is no requirement to "flip" one of the functions about the origin, as is necessary in the general convolution procedure. This gives us the following important fact:

$f(\tau)$ *is just the area under the product of* $h(t)$ *and* $h(t - \tau)$.

The term "area" in this statement refers to integration with respect to the variable t (not τ).

Next, we show that f is an *even* function; that is, it is symmetrical about the origin: $f(u) = f(-u)$ for all real numbers u. The fact that f is a symmetrical function is useful, both for the evaluation of the function and for the subsequent convolution of f with the input autocorrelation function to find the output autocorrelation function. To show that f is symmetrical, we begin with

$$f(-u) = \int_{-\infty}^{\infty} \tilde{h}(-u - v)h(v)\, dv$$

$$= \int_{-\infty}^{\infty} h(u + v)h(v)\, dv.$$

The change of variable $\lambda = u + v$ gives

$$f(-u) = \int_{-\infty}^{\infty} h(\lambda)h(\lambda - u)\, d\lambda$$

$$= \int_{-\infty}^{\infty} h(\lambda - u)h(\lambda)\, d\lambda.$$

A comparison with the last integral and (3.52) shows that this is just $f(u)$. Hence, we have established the relationship

$$f(-u) = f(u).$$

One consequence of this relationship is that it suffices to evaluate the integral in (3.53) for $\tau \geq 0$ only. We then set $f(\tau) = f(-\tau)$ for $\tau < 0$.

If $h(t)$ happens to be time limited, the integral in (3.53) is equivalent to an integral with *finite* limits, as illustrated by the next example.

Example 3–4   A time-limited impulse response

Suppose T is a (finite) positive number and the impulse response h satisfies $h(t) = 0$ for all values of t that are not in the range $0 \leq t \leq T$. Such an impulse response is said to be *time limited—in this case to the interval* $[0, T]$. For an impulse response that is time limited to $[0, T]$, $h(t - \tau) = 0$ for $t < \tau$ and $h(t - \tau) = 0$ for $t > T + \tau$. If $\tau > 0$, the product $h(t)h(t - \tau)$ is zero for all t outside the range $\tau \leq t \leq T$, as can be seen from Figure 3–3. So if h is time limited to $[0, T]$ and $0 < \tau \leq T$, then (3.53) reduces to

$$f(\tau) = \int_{\tau}^{T} h(t)h(t - \tau)\, dt.$$

Notice that, because the impulse response is time limited to $[0, T]$, $f(\tau) = 0$ for $\tau > T$. ∎

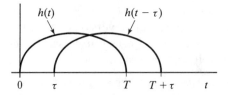

FIGURE 3–3   An illustration of $h(t)$ and $h(t - \tau)$ if h is time limited to $[0, T]$.

The procedure for evaluating the function f is illustrated in the next three exercises. Analytical methods and graphical methods can be used in this evaluation. In Exercise 3–5, a purely anaytical derivation is given for a specific example. For most impulse responses, however, it is best to use a combination of analytical and graphical methods in the evaluation of f. That is, we use sketches of the functions, such as those in Figure 3–3, to help set up the necessary integrals. For very simple functions, graphical methods alone suffice, as is illustrated later by Exercise 3–6.

Exercise 3–5. The rectangular pulse of duration $T > 0$ is defined by $p_T(t) = 1$ for $0 \leq t < T$ and $p_T(t) = 0$ otherwise. Using analytical methods, determine the function $f = \widetilde{h} * h$ for the linear filter with impulse response $h(t) = p_T(t)$.

Solution. The impulse response is time limited to $[0, T]$, so the expression developed in Example 3–4 applies. That is, for $h(t) = p_T(t)$ and $0 < \tau < T$,

$$f(\tau) = \int_\tau^T h(t)h(t - \tau)\, dt = \int_\tau^T p_T(t)p_T(t - \tau)\, dt.$$

As mentioned in that example, $f(\tau) = 0$ for $\tau > T$. The product $p_T(t)p_T(t - \tau)$ is zero for a given value of t and τ if *either* pulse function takes the value zero for that t and τ. The product is nonzero only if both pulse functions are nonzero, in which case the product is unity. Now, $p_T(t)$ is nonzero for $0 \leq t < T$, and $p_T(t - \tau)$ is nonzero for $\tau \leq t < \tau + T$, so the product is nonzero for $\tau \leq t < T$. That is, for $0 < \tau < T$, $p_T(t)p_T(t - \tau) = 1$ for all t in the range $\tau \leq t < T$. It follows that

$$f(\tau) = \int_\tau^T 1\, dt = T - \tau, \quad \text{for} \quad 0 < \tau < T.$$

We have already established that $f(\tau) = 0$ for $\tau > T$ (Example 3–4). Next, we use the symmetry of the function f to conclude that, $f(\tau) = 0$ for $\tau < -T$ and $f(\tau) = T + \tau$ for $-T \leq \tau < 0$. Thus,

$$f(\tau) = 0 \quad \text{for} \quad |\tau| \geq T$$

and

$$f(\tau) = T - |\tau| \quad \text{for} \quad |\tau| < T.$$

$f(\tau)$ is the triangular function illustrated in Figure 3–4. ■

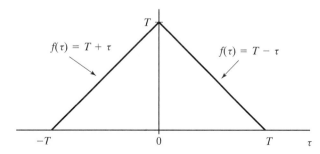

FIGURE 3–4 The function $f = \widetilde{h} * h$ for $h(t) = p_T(t)$.

Exercise 3–6. Using graphical methods, determine the function $f = \tilde{h} * h$ for the linear filter with impulse response $h(t) = p_T(t)$.

Solution. The solution presented here is unfair in one respect: We use the fact that f is an even function, which was proved analytically. However, the reader should be able to make a few sketches to provide a convincing "graphical proof" that f is an even function. Given that f is an even function, we can restrict attention to $\tau \geq 0$. First, we sketch $h(t), h(t - \tau)$, and the product $h(t)h(t - \tau)$, as shown in Figure 3–5. Next, we determine the area under the product. As can be seen in the figure, this area is just the area of a rectangle of width $T - \tau$ and unity height, provided that $\tau < T$. Thus, we see that $f(\tau) = T - \tau$ for $0 \leq \tau < T$. Inspection of the graphs of $h(t)$ and $h(t - \tau)$ shows that the product $h(t)h(t - \tau)$ is identically zero if $\tau > T$. This takes care of all nonnegative values of τ. The values of $f(\tau)$ for negative values of τ are determined from the fact that f is an even function. ■

Exercise 3–7. Evaluate the function $f = \tilde{h} * h$ for a time-invariant linear system with impulse response $h(t) = \beta \exp(-\alpha t)$ for $t \geq 0$ and $h(t) = 0$ for $t < 0$. (Assume that α and β are positive constants.)

Solution. First observe that the impulse response can be written as

$$h(t) = \beta \exp(-\alpha t)u(t),$$

where $u(\cdot)$ is the unit step function (i.e., $u(\lambda) = 1$ for $\lambda \geq 0$ and $u(\lambda) = 0$ for $\lambda < 0$). Note that h is not time limited, so the expression developed in Example 3–4 does not apply. From the more general expression in (3.53), we see that the function $f = \tilde{h} * h$ is given by

$$f(\tau) = \int_{-\infty}^{\infty} h(t)h(t - \tau)\, dt$$

$$= \beta^2 \int_{-\infty}^{\infty} e^{-\alpha t}u(t)e^{-\alpha(t - \tau)}u(t - \tau)\, dt.$$

Illustrations of the functions appearing in the integrands of these two integrals are given in Figure 3–6 for $\tau \geq 0$.

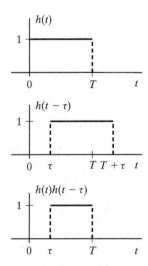

FIGURE 3–5   Illustration of the graphical method for convolving two rectangular pulses.

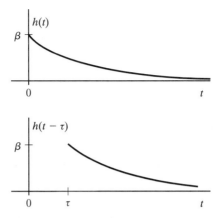

FIGURE 3–6   Original and delayed versions of the impulse response $h(t) = \beta \exp(-\alpha t) u(t)$.

Because we can restrict our attention to $\tau \geq 0$, we see that the preceding integral reduces to

$$f(\tau) = \beta^2 e^{\alpha \tau} \int_\tau^\infty e^{-2\alpha t}\, dt.$$

Evaluating this integral, we find that

$$f(\tau) = \left(\frac{\beta^2}{2\alpha}\right) e^{-\alpha \tau}, \quad \tau \geq 0,$$

which implies that

$$f(\tau) = f(-\tau) = \left(\frac{\beta^2}{2\alpha}\right) e^{+\alpha \tau}, \quad \tau < 0.$$

Thus, $f(\tau) = (\beta^2/2\alpha) \exp(-\alpha|\tau|)$ for $-\infty < \tau < \infty$.   ∎

Next, we examine in greater detail the evaluation of the output autocorrelation function for a time-invariant linear system that has a wide-sense stationary random process as its input. Equations (3.32) and (3.47) imply that the output random process is wide-sense stationary, so the output autocorrelation can be obtained from $R_Y = (\widetilde{h} * h) * R_X$. In working with convolutions, it is important to recall that the convolution of two functions is a *function*, not a number. The convolution of $\widetilde{h}$ and h is the function f defined by (3.53) for each value of its argument. To find R_Y, the function f is convolved with the function R_X. That is,

$$R_Y(\tau) = \int_{-\infty}^\infty f(\tau - u) R_X(u)\, du$$

for each real number τ. Because $f(\tau - u) = f(u - \tau)$, the preceding equation is equivalent to

$$R_Y(\tau) = \int_{-\infty}^\infty R_X(u) f(u - \tau)\, du. \tag{3.54}$$

Just like the integral in (3.53), the integral in (3.54) is simpler than the general convolution integral. All that is needed is to integrate the product of the autocorrelation function and a delayed version of the function f.

The simplest procedure for evaluating the output autocorrelation function for most filter impulse responses and input autocorrelation functions is to first determine the function f and then convolve f with the input autocorrelation function R_X to get the output autocorrelation function. That is, we first evaluate $f = \tilde{h} * h$ and then evaluate $R_Y = f * R_X$. The second part of this procedure can be carried out by applying (3.54), whose evaluation is simplified further by the fact that each of the functions f, R_X, and R_Y is even. Earlier in this section, we proved that f is even, and we proved in Chapter 2 that all autocorrelation functions for wide-sense stationary processes are even, so both R_X and R_Y are even. This implies that (3.54) need be evaluated for $\tau \geq 0$ only. The values of $R_Y(\tau)$ for $\tau < 0$ can be obtained from $R_Y(\tau) = R_Y(-\tau)$. The procedure for evaluating $R_Y = f * R_X$ via (3.54) is illustrated in the next exercise.

Exercise 3–8. Consider the linear system of Exercise 3–5 with impulse response $h(t) = p_T(t)$. The input is a wide-sense stationary random process $X(t)$ with autocorrelation function

$$R_X(\tau) = \sigma^2 \exp(-\gamma|\tau|), \quad -\infty < \tau < \infty,$$

where γ is a positive constant. Find the autocorrelation function for the output process $Y(t)$.

Solution. In Exercise 3–5, the function f is shown to be the triangular function $f(\tau) = T - |\tau|$ for $|\tau| < T$ and $f(\tau) = 0$ otherwise. All that remains is to evaluate the convolution $R_Y = f * R_X$. From (3.54), we see that the corresponding convolution integral is equivalent to

$$R_Y(\tau) = \int_{-\infty}^{\infty} \sigma^2 \exp(-\gamma|u|) f(u - \tau) \, du.$$

The functions appearing in the integrand are illustrated in Figure 3–7 for $0 < \tau < T$. Notice that for any $\tau > 0$, the integrand is identically zero on the intervals $(-\infty, \tau - T)$ and $(\tau + T, \infty)$. Therefore, for $\tau > 0$, we have

$$R_Y(\tau) = \int_{\tau-T}^{\tau+T} \sigma^2 \exp(-\gamma|u|) f(u - \tau) \, du$$

$$= \int_{\tau-T}^{\tau+T} \sigma^2 \exp(-\gamma|u|) \{T - |u - \tau|\} \, du$$

$$= \int_{\tau-T}^{\tau+T} \sigma^2 T \exp(-\gamma|u|) \, du - \int_{\tau-T}^{\tau+T} \sigma^2 |u - \tau| \exp(-\gamma|u|) \, du.$$

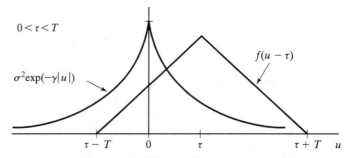

$0 < \tau < T$

$\sigma^2 \exp(-\gamma|u|)$

$f(u - \tau)$

$\tau - T \qquad 0 \qquad \tau \qquad \tau + T \quad u$

FIGURE 3–7   The functions $\sigma^2 \exp(-\gamma|u|)$ and $f(u - \tau)$ for $0 < \tau < T$.

Next, for $0 < \tau < T$, we write the first of these two integrals as the sum of two integrals and the second as the sum of three integrals. (This part will be different for $\tau \geq T$). The regions of integration for the new integrals are regions over which the functional forms of the integrands are constant. The term $|u - \tau|$ changes form (from a positive slope to a negative slope) at $u = \tau$, and the term $\exp(-\gamma|u|)$ changes form at $u = 0$.

Therefore, for $0 < \tau < T$, the foregoing expression for $R_Y(\tau)$ is equivalent to

$$
R_Y(\tau) = \sigma^2 T \int_{\tau-T}^{0} \exp(+\gamma u)\, du + \sigma^2 T \int_{0}^{\tau+T} \exp(-\gamma u)\, du
$$

$$
- \sigma^2 \int_{\tau-T}^{0} (-u + \tau)\exp(+\gamma u)\, du - \sigma^2 \int_{0}^{\tau} (-u + \tau)\exp(-\gamma u)\, du
$$

$$
- \sigma^2 \int_{0}^{\tau+T} (u - \tau)\exp(-\gamma u)\, du.
$$

All that remains is to evaluate these integrals for τ in the range $0 < \tau < T$, a task that is left to the reader, who should also derive the expression for $R_Y(\tau)$ for $\tau \geq T$. ∎

Frequently, it is necessary to determine the power in the output process from the autocorrelation function of the input process and the impulse response of the linear system. Recall that the expected value of the instantaneous power in the process $Y(t)$ is equal to the second moment $E\{[Y(t)]^2\}$. If the input process $X(t)$ is wide-sense stationary and the linear filter is time invariant, then

$$
E\{[Y(t)]^2\} = R_Y(0).
$$

If all we need to compute is the second moment of the output process, rather than its complete autocorrelation function, the evaluation is considerably simpler. In particular, no convolutions are required once the function f has been obtained. To see this, consider the fact that

$$
R_Y(0) = \int_{-\infty}^{\infty} f(\lambda) R_X(\lambda)\, d\lambda, \tag{3.55a}
$$

which follows from (3.54) with $\tau = 0$. Note that if $X(t)$ is *white noise*, $R_X(\tau) = c\delta(\tau)$ for some positive constant c, and thus,

$$
R_Y(0) = cf(0) = c \int_{-\infty}^{\infty} [h(t)]^2\, dt.
$$

From (3.55a), we see that the second moment of the output random process is just the integral of the product of the functions f and R_X. Because these functions are both *even*, it follows that if f and R_X have no delta functions at the origin, then

$$
R_Y(0) = 2 \int_{0}^{\infty} f(\lambda) R_X(\lambda)\, d\lambda. \tag{3.55b}
$$

Applications of this result are illustrated in the next two exercises.

Exercise 3–9. Suppose the random process $X(t)$ has zero mean and autocorrelation function

$$
R_X(\tau) = \begin{cases} (\tau_0 - |\tau|)/\tau_0, & 0 \leq |\tau| \leq \tau_0, \\ 0, & |\tau| > \tau_0. \end{cases}
$$

Let $X(t)$ be the input to a time-invariant linear system with impulse response

$$h(t) = \begin{cases} 1, & 0 \leq t < t_0, \\ 0, & \text{otherwise}, \end{cases}$$

where $\tau_0 > t_0$. Find the power in the output process $Y(t)$.

Solution. The process is wide-sense stationary, so its power is $R_Y(0)$. To find $R_Y(0)$, we first determine the function f, which is related to the impulse response of the filter by $f = \tilde{h} * h$. Thus, to find the function f, we must convolve two rectangular pulses, each of which has duration t_0. But the convolution of any two rectangular pulses of duration t_0 is a triangular pulse of duration $2t_0$, so $\tilde{h} * h$ is such a triangular pulse. In particular, because we know that $f = \tilde{h} * h$ is a symmetrical function, the triangular pulse must be centered at the origin. In addition, the height of the triangle is equal to $f(0)$, which is just the area under the square of the impulse response. [To see this, let $\tau = 0$ in (3.53).] For the given impulse response, the said area is t_0. Putting all these facts together, we conclude that $f(t)$ is a triangular pulse of duration $2t_0$, the peak of the triangle must be at the origin, and $f(0) = t_0$. Thus, the function

$$f(t) = \begin{cases} t_0 - |t|, & -t_0 \leq t \leq t_0, \\ 0, & |t| > t_0. \end{cases}$$

Given the function f, we can find $R_Y(0)$ by substituting into (3.55b) for $f(\lambda)$ to obtain

$$R_Y(0) = 2 \int_0^{t_0} (t_0 - \lambda) R_X(\lambda)\, d\lambda.$$

Because $\tau_0 > t_0$, this equation is equivalent to

$$R_Y(0) = 2\tau_0^{-1} \int_0^{t_0} (t_0 - \lambda)(\tau_0 - \lambda)\, d\lambda = t_0^2 (3\tau_0)^{-1}[3\tau_0 - t_0].$$

If t_0 were greater than τ_0, the upper limit on the integral would be τ_0 (i.e., in general, the upper limit is the minimum of t_0 and τ_0). ∎

Exercise 3–10. Let $X(t)$ be a zero-mean wide-sense stationary process. Define a random process $Z(t), t \geq 0$, by the mean-square integral

$$Z(t) = \int_0^t X(\tau)\, d\tau.$$

Find the variance of $Z(t)$ in terms of the autocorrelation function for the process $X(t)$.

Solution. Consider an arbitrary $t_0 > 0$ and evaluate $\text{Var}\{Z(t_0)\}$. First, we define an impulse response h for a hypothetical linear time-invariant system by

$$h(\tau) = \begin{cases} 1, & 0 \leq \tau \leq t_0, \\ 0, & \text{otherwise}. \end{cases}$$

Next, notice that since $0 \leq t_0 - \tau \leq t_0$ if and only if $0 \leq \tau \leq t_0$, then

$$h(\tau) = h(t_0 - \tau) = \begin{cases} 1, & 0 \leq t_0 - \tau \leq t_0, \\ 0, & \text{otherwise}. \end{cases}$$

The reason we are interested in the impulse response $h(t)$ is the fact that

$$Z(t_0) = \int_0^{t_0} X(\tau)\, d\tau = \int_{-\infty}^{\infty} h(\tau) X(\tau)\, d\tau = \int_{-\infty}^{\infty} h(t_0 - \tau) X(\tau)\, d\tau,$$

for $t_0 > 0$. That is, if $X(t)$ is the input to this filter, $Z(t_0)$ can be obtained by sampling the output of the filter at time t_0. If we define the random process $Y(t)$, $-\infty < t < \infty$, by

$$Y(t) = \int_{-\infty}^{\infty} h(t_0 - \tau) X(\tau)\, d\tau,$$

then $Z(t_0) = Y(t_0)$, and $Y(t)$ is wide-sense stationary even though $Z(t)$ is not.

From the preceding results, we see that the mean of $Z(t_0)$ is given by

$$\mu_Z(t_0) = \mu_Y = \mu_X \int_{-\infty}^{\infty} h(t_0 - \tau)\, d\tau = 0,$$

since $\mu_X = 0$. Furthermore,

$$\mathrm{Var}\{Z(t_0)\} = \mathrm{Var}\{Y(t_0)\} = R_Y(t_0, t_0) = R_Y(0).$$

From (3.55) and Exercise 3–9, we see that for this particular impulse response h,

$$R_Y(0) = \int_{-t_0}^{t_0} (t_0 - |\lambda|) R_X(\lambda)\, d\lambda = 2 \int_0^{t_0} (t_0 - \lambda) R_X(\lambda)\, d\lambda,$$

where the first integral is valid in general, but the second integral requires that R_X not have a delta function at the origin.

Because t_0 is arbitrary, we conclude from the foregoing results that, for each $t > 0$,

$$\mathrm{Var}\{Z(t)\} = \int_{-t}^{t} \left(t - |\lambda| \right) R_X(\lambda)\, d\lambda = 2 \int_0^{t} (t - \lambda) R_X(\lambda)\, d\lambda, \qquad (3.56)$$

with the same restriction that there must not be a delta function at the origin in the rightmost integral.

As an example, suppose $X(t)$ has the autocorrelation function given by

$$R_X(\tau) = \sigma^2 \exp\left(-\gamma|\tau|\right), \quad -\infty < \tau < \infty,$$

which is examined in Exercise 3–8. For this autocorrelation function,

$$\mathrm{Var}\{Z(t)\} = 2 \int_0^{t} (t - \lambda) \sigma^2 e^{-\gamma\lambda}\, d\lambda$$

$$= 2\sigma^2 \gamma^{-2}[\gamma t - 1 + \exp(-\gamma t)]$$

for $t > 0$.

An alternative approach is to work directly with the definition of the process $Y(t)$ as follows: For $t > 0$ and $s > 0$,

$$R_Z(t, s) = E\{Z(t)Z(s)\}$$

$$= E\left\{ \int_0^{t} X(\tau)\, d\tau \int_0^{s} X(\lambda)\, d\lambda \right\} = \int_0^{t} \int_0^{s} E\{X(\tau)X(\lambda)\}\, d\lambda\, d\tau$$

$$= \int_0^{t} \int_0^{s} R_X(\tau - \lambda)\, d\lambda\, d\tau.$$

As a special case, for $t > 0$,

$$\text{Var}\{Z(t)\} = R_Z(t, t) = \int_0^t \int_0^t R_X(\tau - \lambda) \, d\lambda \, d\tau,$$

which can be shown to reduce to (3.56) by the change of variable $u = \tau, v = \tau - \lambda$. (This change of variable is illustrated in detail in Section 4.3.) ∎

Before proceeding, it is beneficial to summarize certain key results obtained so far in this subsection. These results provide the methods for evaluating the correlation functions that arise in the analysis of random processes in continuous-time linear systems. First, for *general* second-order continuous-time random processes and linear systems, the primary results are

$$R_{Y,X}(t, s) = \int_{-\infty}^{\infty} h(t, \tau) R_X(\tau, s) \, d\tau$$

for the crosscorrelation between the input and output random processes and

$$R_Y(t, s) = \int_{-\infty}^{\infty} \int_{-\infty}^{\infty} h(s, \lambda) h(t, \tau) R_X(\tau, \lambda) \, d\tau \, d\lambda$$

for the autocorrelation function for the output random process. For the special case of continuous-time *wide-sense stationary* random processes and *time-invariant* linear systems, these results simplify to give

$$R_{Y,X}(\tau) = \int_{-\infty}^{\infty} h(\tau - \alpha) R_X(\alpha) \, d\alpha$$

$$= \int_{-\infty}^{\infty} h(\alpha) R_X(\tau - \alpha) \, d\alpha$$

and

$$R_Y(\tau) = \int_{-\infty}^{\infty} h(\beta) \left\{ \int_{-\infty}^{\infty} h(\alpha) R_X[(\tau + \beta) - \alpha] \, d\alpha \right\} d\beta.$$

These last two results can be written in terms of convolutions: $R_{Y,X} = h * R_X$ and $R_Y = \widetilde{h} * h * R_X$. In most cases, the simplest procedure for evaluating the autocorrelation function R_Y is to find $f = \widetilde{h} * h$ and then use $R_Y = f * R_X$. Each of the functions involved in this last convolution is an even function. In order to find the function f, it is usually simpler to use the alternative expression

$$f(\tau) = \int_{-\infty}^{\infty} h(t) h(t - \tau) \, dt,$$

which expresses $f(\tau)$ as the integral of the product of the impulse response and a delayed version of itself. The amount of the delay is τ, the argument of the function. This integral is simpler than the general convolution integral.

Analogous equations can be obtained for linear *discrete-time* systems. Generally, these equations correspond to the results for continuous-time systems if the integrals are replaced by sums. For the results that follow, we assume that (3.21) is satisfied. First, for general discrete-time random processes and linear systems,

$$R_{Y,X}(k,i) = \sum_{n=-\infty}^{\infty} h(k,n) R_X(n,i) \tag{3.57}$$

and

$$R_Y(k,i) = \sum_{m=-\infty}^{\infty} h(i,m) R_{Y,X}(k,m), \tag{3.58}$$

or

$$R_Y(k,i) = \sum_{m=-\infty}^{\infty} \sum_{n=-\infty}^{\infty} h(i,m) h(k,n) R_X(n,m). \tag{3.59}$$

If the input process is wide-sense stationary and the linear system is time invariant, then the output process is also wide-sense stationary and has autocorrelation function

$$R_Y(k) = \sum_{j=-\infty}^{\infty} h(j) R_{Y,X}(k+j), \tag{3.60}$$

or

$$R_Y(k) = \sum_{n=-\infty}^{\infty} \tilde{h}(n) R_{Y,X}(k-n), \tag{3.61}$$

where $\tilde{h}(n) = h(-n)$ for each n and

$$R_{Y,X}(i) = \sum_{m=-\infty}^{\infty} h(m) R_X(i-m). \tag{3.62}$$

That is,

$$R_Y = \tilde{h} * h * R_X, \tag{3.63}$$

where the convolutions are now *discrete* and h is the pulse response of the discrete-time linear time-invariant system. Equations (3.60) and (3.62) can be combined to give a single equation for R_Y in terms of R_X. This equation,

$$R_Y(k) = \sum_{j=-\infty}^{\infty} \sum_{m=-\infty}^{\infty} h(j) h(m) R_X(k+j-m), \tag{3.64}$$

is just the discrete-time counterpart to (3.49). However, for the evaluation of R_Y, it is easier in most cases to evaluate $f = \tilde{h} * h$ from

$$f(i) = \sum_{n=-\infty}^{\infty} \tilde{h}(i-n) h(n)$$

$$= \sum_{n=-\infty}^{\infty} h(n-i) h(n) \tag{3.65}$$

and then evaluate $R_Y = f * R_X$ from

$$R_Y(k) = \sum_{n=-\infty}^{\infty} f(k-n) R_X(n). \tag{3.66}$$

We have already given one exercise (Exercise 3–1) that illustrates the evaluation of the output autocorrelation function for a discrete-time linear system. Other examples are provided by the exercises that follow.

Exercise 3–11. Consider the discrete-time linear system described in Exercise 3–4. Let the input process $X(k)$ have autocorrelation function $R_X(k) = \sigma^2$ for $k = 0$ and $R_X(k) = 0$ for $k \neq 0$. Find the output autocorrelation function and the input–output crosscorrelation function.

Solution. It was shown in the solution to Exercise 3–4 that the system is time invariant with impulse response

$$h(n) = \begin{cases} 2^{-n}, & n \geq 0, \\ 0, & n < 0. \end{cases}$$

Therefore, from (3.62),

$$R_{Y,X}(i) = \sum_{m=0}^{\infty} 2^{-m} R_X(i - m) = \begin{cases} 2^{-i}\sigma^2, & i \geq 0 \\ 0, & i < 0 \end{cases}.$$

We can then use (3.60) to see that

$$R_Y(k) = \sum_{j=0}^{\infty} 2^{-j} R_{Y,X}(k + j).$$

For $k \geq 0$,

$$R_Y(k) = \sum_{j=0}^{\infty} 2^{-j} 2^{-(k+j)} \sigma^2 = \sigma^2 2^{-k} \sum_{j=0}^{\infty} 4^{-j}$$

$$= \left(\tfrac{4}{3}\right)\sigma^2 2^{-k},$$

and for $k < 0$,

$$R_Y(k) = R_Y(-k) = \left(\tfrac{4}{3}\right)\sigma^2 2^k.$$

Therefore, for each k,

$$R_Y(k) = \left(\tfrac{4}{3}\right)\sigma^2 2^{-|k|}.$$

An alternative approach is to first find $f = \tilde{h} * h$ and then determine $R_Y = f * R_X$. Using (3.65), we see that

$$f(i) = \sum_{n=0}^{\infty} h(n - i)2^{-n} = \left(\tfrac{4}{3}\right)2^{-|i|}.$$

Then, (3.66) gives

$$R_Y(k) = \sum_{n=-\infty}^{\infty} f(k - n)R_X(n) = f(k)R_X(0) = f(k)\sigma^2$$

$$= \left(\tfrac{4}{3}\right)\sigma^2 2^{-|k|}. \qquad \blacksquare$$

Exercise 3–12. Consider a discrete-time system for which the output consists of differences between successive values of the input. That is, if x is the input signal, the output is

$$y(k) = x(k) - x(k - 1). \tag{3.67}$$

Suppose the input is a discrete-time wide-sense stationary random process with autocorrelation function

$$R_X(n) = \sigma^2 e^{-\beta|n|}, \tag{3.68}$$

where $\beta > 0$. Find the autocorrelation function of the output.

Solution. Two approaches are presented.

(a) For input process $X(k)$, the output process is given by

$$Y(k) = X(k) - X(k-1)$$

for all k. Therefore, from $R_Y(k, i) = E\{Y(k)Y(i)\}$, it follows that

$$
\begin{aligned}
R_Y(k, i) &= E\{[X(k) - X(k-1)][X(i) - X(i-1)]\} \\
&= R_X(k-i) - R_X(k-i+1) - R_X(k-1-i) + R_X(k-i).
\end{aligned}
$$

Thus, $R_Y(k, i)$ depends on $k - i$ only. That is,

$$
\begin{aligned}
R_Y(k, i) &= R_Y(k-i, 0) = R_Y(k-i) \\
&= 2R_X(k-i) - R_X(k-i+1) - R_X(k-i-1) \\
&= \sigma^2\{2e^{-\beta|k-i|} - e^{-\beta|k-i+1|} - e^{-\beta|k-i-1|}\}.
\end{aligned}
$$

This equation can be simplified somewhat by writing it as

$$R_Y(n) = \sigma^2\{2e^{-\beta|n|} - e^{-\beta|n+1|} - e^{-\beta|n-1|}\}. \tag{3.69}$$

Next, observe that, for $n > 0$,

$$|n + 1| = |n| + 1$$

and

$$|n - 1| = |n| - 1,$$

and for $n < 0$,

$$|n + 1| = |n| - 1 \quad \text{and} \quad |n - 1| = |n| + 1.$$

In either case,

$$e^{-\beta|n+1|} + e^{-\beta|n-1|} = e^{-\beta|n|}\{e^{-\beta} + e^{\beta}\}$$

(for $n \neq 0$). Thus,

$$R_Y(n) = \begin{cases} 2\sigma^2\{1 - e^{-\beta}\}, & n = 0, \\ \sigma^2\{2 - (e^{-\beta} + e^{\beta})\}\exp\{-\beta|n|\}, & n \neq 0. \end{cases}$$

(b) For the *second* approach, we first observe that the system described by (3.67) has pulse response

$$h(n) = \begin{cases} 1, & n = 0, \\ -1, & n = 1, \\ 0, & \text{otherwise}. \end{cases}$$

We then use (3.65) to conclude that $f = \tilde{h} * h$ is given by

$$
f(i) = \begin{cases} 2, & i = 0, \\ -1, & |i| = 1, \\ 0, & \text{otherwise.} \end{cases}
$$

Hence, (3.66) implies that

$$
R_Y(k) = \sum_{n=k-1}^{k+1} f(k - n)R_X(n)
$$

$$
= -R_X(k - 1) + 2R_X(k) - R_X(k + 1).
$$

Using (3.68) to substitute into the preceding expression gives (3.69), which can then be simplified, as in part (a). ∎

Exercise 3–13. Show that the inequality

$$
\sum_{i=-\infty}^{\infty} |h(k, i)||\mu_X(i)| < \infty
$$

(which is (3.38)) is satisfied whenever (3.9) and (3.22) hold. Thus, if the discrete-time linear system is stable and the second moment of the random process $X(k)$ is bounded, then the mean of the random process $Y(k)$ is finite.

Solution. We use the fact that $|x| \leq x^2 + 1$ for any real number x (see the solution to Exercise 2–6) to obtain

$$
|\mu_X(i)| \leq [\mu_X(i)]^2 + 1 \leq \text{Var}\{X(i)\} + [\mu_X(i)]^2 + 1.
$$

Since $\text{Var}\{X(i)\} + [\mu_X(i)]^2 = R_X(i, i)$, the preceding inequality and (3.22) imply

$$
|\mu_X(i)| \leq R_X(i, i) + 1 \leq r + 1. \tag{3.70}
$$

Combining (3.70) and (3.9) we have

$$
\sum_{i=-\infty}^{\infty} |h(k, i)||\mu_X(i)| \leq (r + 1) \sum_{i=-\infty}^{\infty} |h(k, i)| < \infty,
$$

which establishes the desired inequality and proves that the mean of $Y(k)$ is finite.

It should be clear at this point that the same arguments can be applied to continuous-time processes and systems to show that (3.6) and (3.26) imply (3.34), which in turn implies that the mean of the output of the continuous-time system is finite. ∎

On the basis of the results of the analysis of this section, we can state some general properties of *time-invariant* linear systems with inputs that are second-order random processes. These properties hold for both discrete- and continuous-time systems and processes. We assume, of course, that the system pulse or impulse response is known and satisfies conditions required to guarantee the output is a second-order random process [e.g., (3.21) or (3.25)].

1. *The output autocorrelation function can be determined from knowledge of the input autocorrelation function only.*

2. *The crosscorrelation function for the input and output processes can be determined from knowledge of the input autocorrelation function only.*

3. *If the input process is wide-sense stationary, the input and output processes are jointly wide-sense stationary, and hence, the output process is wide-sense stationary.*

4. *If the input process is covariance stationary, the output process is also covariance stationary.*

Properties 1 and 2 are also true for time-varying linear systems, but properties 3 and 4 depend critically on the time invariance of the system.

3.4 GAUSSIAN RANDOM PROCESSES IN LINEAR SYSTEMS

It is pointed out in Chapter 2 that the mean and autocorrelation functions completely specify the finite-dimensional distributions for a *Gaussian* random process. (This is *not* true for general random processes.) Thus, if we know the mean and autocorrelation functions for a Gaussian random process $X(t)$, we can, in principle, compute the probabilities of any events involving a finite number of samples of $X(t)$. If $X(t)$ is a Gaussian random process, it suffices to know the mean and autocorrelation function for $X(t)$ in order to compute such probabilities as the following:

1. the probability that $X(t_0)$ is negative;

2. the probability that $X(t_1) + X(t_2)$ is less than 2;

3. the probability that $[X(t_1) + X(t_2)]^2$ is between 10 and 20;

4. the probability that $X(t_1) + X(t_2) + \ldots + X(t_{100})$ is greater than some threshold value γ.

All of these examples involve a finite number of random variables only; hence, the finite-dimensional distributions for the random process $X(t)$ provide enough information. The first example requires knowledge of the one-dimensional distribution only, and the second and third require knowledge of the two-dimensional distribution only, but the fourth example requires knowledge of the 100-dimensional distribution. Fortunately, because $X(t)$ is a Gaussian random process, all of these distributions can be determined from the mean and autocorrelation function for $X(t)$.

—In many cases, we do not have to actually write down the higher dimensional distributions in order to solve the problem. If $X(t)$ is a Gaussian random process, then any linear combination of samples of $X(t)$ is a Gaussian random variable. That is, given the sampling times $t_1, t_2, \ldots, t_n$ and the coefficients $\alpha_1, \alpha_2, \ldots, \alpha_n$, the random variable

$$Z = \sum_{i=0}^{n} \alpha_i X(t_i) \tag{3.71}$$

is a Gaussian random variable. So, for instance,

$$P(Z \le u) = \Phi\left(\frac{u - \mu}{\sigma}\right),$$

where $\mu = E\{Z\}$ and $\sigma = \sqrt{\text{Var}\{Z\}}$. Thus, it suffices to find the mean and variance of Z by using (3.71). Of course, if we wish to find $P(Z \ge u)$, rather than $P(Z \le u)$, we can use the fact that

$$P(Z \ge u) = 1 - P(Z < u) = 1 - [P(Z \le u) - P(Z = u)],$$

and because the Gaussian distribution function is continuous, $P(Z = u) = 0$ for each u. It follows that if Z is Gaussian, then

$$P(Z \ge u) = 1 - P(Z \le u) = 1 - \Phi\left(\frac{u - \mu}{\sigma}\right).$$

Similarly,

$$P(v \le Z \le u) = P(Z \le u) - P(Z < v),$$

and because Z is Gaussian,

$$P(v \le Z \le u) = \Phi\left(\frac{u - \mu}{\sigma}\right) - \Phi\left(\frac{v - \mu}{\sigma}\right).$$

The general procedure just illustrated permits the computation of probabilities such as the foregoing 1–4. However, with number 3, do not make the unforgivable mistake of assuming that $[X(t_1) + X(t_2)]^2$ is Gaussian. Although it is true that $Z = X(t_1) + X(t_2)$ is Gaussian, *the square of a Gaussian random variable is definitely not Gaussian*. However, we can use the fact that for $0 < v < u$,

$$P(v \le Z^2 \le u) = P(-\sqrt{u} \le Z \le -\sqrt{v}) + P(\sqrt{v} \le Z \le \sqrt{u}).$$

If Z is a continuous random variable, this is valid for $v = 0$ as well. Next, the fact that Z is Gaussian can be used to express $P(v \le Z^2 \le u)$ in terms of the function Φ.

Examples such as number 4 in the list do not arise often in practice, and even when they do, it is rarely, if ever, necessary to determine the 100-dimensional density or distribution function in order to carry out the calculation. Normally, one should determine the mean μ and variance σ^2 of the sum $W = X(t_1) + X(t_2) + \ldots + X(t_{100})$ and then use the fact that a linear combination of jointly Gaussian random variables is Gaussian.

Hence, W is a Gaussian random variable with distribution function

$$F_W(u) = P(W \le u) = \Phi\left(\frac{u - \mu}{\sigma}\right),$$

so

$$P(W > \gamma) = 1 - P(W \le \gamma) = 1 - F_W(\gamma) = 1 - \Phi\left(\frac{\gamma - \mu}{\sigma}\right).$$

Now, μ, the mean of the sum, is just the sum of the means, but, in general, the variance of the sum is not just the sum of the variances. (Why?) Nevertheless, σ^2 can be determined from the autocorrelation function of the random process $X(t)$ with the use of standard results on sums of random variables (e.g., see the solution to Exercise 3–1).

The preceding examples involve functions of a finite number of samples of the random process $X(t)$. In many engineering investigations, such as the study of noise in electronic systems, it is necessary to be able to work with functions of an infinite number of samples of $X(t)$. In fact, the number of samples may be uncountably infinite; for example, we may have to deal with random variables and random processes that depend on $X(t)$ as t ranges over the set of all real numbers. This is the case, for example, if the random process $Y(t)$, $-\infty < t < \infty$, is the output of a linear system with impulse response h when the input is the random process $X(t)$, $-\infty < t < \infty$; that is, when

$$Y(t) = \int_{-\infty}^{\infty} h(t, \tau) X(\tau) \, d\tau.$$

In general, questions concerning the random process $Y(t)$, or even a finite number of samples of $Y(t)$, cannot be answered if our knowledge of the input random process $X(t)$ is limited to its mean and autocorrelation functions. If $X(t)$ is not Gaussian, then, in general, knowledge of $\mu_X(t)$ and $R_X(t, s)$ for all t and s is not sufficient to determine the distribution of the input process $X(t)$, let alone the distribution of the output process. As we have already demonstrated in Section 3.3, at least the mean and autocorrelation functions for the output process $Y(t)$ can be determined, even if our knowledge of the input $X(t)$ is limited to its mean and autocorrelation functions. Specifically, the output mean can be determined from the input mean, and the output autocorrelation function can be determined from the input autocorrelation function.

For the important special case in which $X(t)$ is a Gaussian random process, we can determine a great deal more about the output than its mean and autocorrelation functions. Suppose $X(t)$, the input to a linear system, is a *Gaussian* random process. Given *only* the mean and autocorrelation functions for $X(t)$, can the finite-dimensional distributions for the output process $Y(t)$ be determined? Fortunately, the answer is that they can. This is because the class of Gaussian random processes is closed under linear operations: *A linear operation on a Gaussian process produces another Gaussian process.* Thus, if the input to a linear filter is a Gaussian random process $X(t)$ with mean $\mu_X(t)$ and autocorrelation $R_X(t, s)$, then the output is a Gaussian random process $Y(t)$ with mean $\mu_Y(t)$ and autocorrelation $R_Y(t, s)$. This information completely specifies the finite-dimensional distributions for the output process. One consequence is that $Y(t)$ is stationary whenever it is wide-sense stationary.

To summarize, for a linear system with given impulse response (in continuous time) or pulse response (in discrete time), knowledge of the input mean and auto-correlation functions is sufficient to determine the finite-dimensional distributions for the output process *if* the input is a *Gaussian* random process. In principle, we can therefore compute the probabilities of any events involving a finite number of samples of $Y(t)$ from the mean and autocorrelation functions for $X(t)$ and the mathematical characterization of the linear system. In most situations, the easiest way to compute such probabilities is to first determine the mean and autocorrelation functions for $Y(t)$.

Recall from Section 3.3 that for continuous-time processes and systems, the output mean and autocorrelation functions are given by

$$\mu_Y(t) = \int_{-\infty}^{\infty} h(t,\tau)\mu_X(\tau)\,d\tau$$

and

$$R_Y(t,s) = \int_{-\infty}^{\infty}\int_{-\infty}^{\infty} h(s,\lambda)h(t,\tau)R_X(\tau,\lambda)\,d\tau\,d\lambda,$$

respectively. For discrete-time processes and systems, these functions are given by

$$\mu_Y(k) = \sum_{i=-\infty}^{\infty} h(k,i)\mu_X(i)$$

and

$$R_Y(k,i) = \sum_{m=-\infty}^{\infty}\sum_{n=-\infty}^{\infty} h(i,m)h(k,n)R_X(n,m).$$

All of these equations have simpler forms, as presented in Section 3.3, if the input random process is wide-sense stationary or if the linear filter is time invariant (or both).

The very important fact that the output of a linear system is a Gaussian random process whenever the input random process is Gaussian is stated in nearly every engineering text on random processes. However, the proof of this result is rarely given in such texts. In fact, the proof is quite straightforward, although it does make use of certain convergence properties of sequences of random variables that have not been discussed in this book. However, it is instructive to at least go through an outline of the proof. In this outline, we will consider only discrete-time random processes and systems.

Recall from Section 3.2 that the output random process $Y(k)$ is related to the input random process by

$$Y_N(k) = \sum_{n=-N}^{N} h(k,n)X(n)$$

and

$$\lim_{N\to\infty} E\{[Y(k) - Y_N(k)]^2\} = 0.$$

That is, $Y(k)$ is a limit (in the mean-square sense) of a finite linear combination of the random variables $X(n), n = 0, \pm 1, \pm 2, \ldots$, provided that this limit exists. To ensure that the mean-square limit does exist, we assume that

$$\sum_{i=-\infty}^{\infty} \sum_{n=-\infty}^{\infty} |h(k,n)h(k,i)R_X(n,i)| < \infty, \quad \text{for each } k.$$

In particular, this implies that the input process must be a second-order process (because it guarantees that $R_X(n,n) < \infty$ for each n).

To prove that the discrete-time process $Y(k)$ is a Gaussian process, we must show that for an arbitrary positive integer m and arbitrary integers $k_1, k_2, \ldots, k_m$, the random variable

$$Z = \sum_{j=1}^{m} a_j Y(k_j) \tag{3.72}$$

has a Gaussian distribution function for each choice of the coefficients $a_1, a_2, \ldots, a_m$. Accordingly, we note that for each j, $Y(k_j)$ is a mean-square limit of a sequence of random variables $Y_N(k_j)$; that is $E\{[Y(k_j) - Y_N(k_j)]^2\} \to 0$ as $N \to \infty$, where

$$Y_N(k_j) = \sum_{n=-N}^{N} h(k_j, n) X(n), \tag{3.73}$$

as in (3.17). Analogously to (3.72), we define

$$Z_N = \sum_{j=1}^{m} a_j Y_N(k_j) \tag{3.74}$$

for each positive integer N. Combining (3.73) and (3.74) we find that

$$Z_N = \sum_{n=-N}^{N} \sum_{j=1}^{m} a_j h(k_j, n) X(n).$$

Since $X(k)$ is a Gaussian random process, this last equation implies that Z_N is a Gaussian random variable for each N.

Next, we note that because each term of the finite sum in (3.74) converges in a mean-square sense to the corresponding term of the sum in (3.72) as $N \to \infty$, Z_N must converge in a mean-square sense to Z as $N \to \infty$. Let G_N be the distribution function for Z_N. Mean-square convergence of Z_N to Z implies that the sequence of distribution functions G_N must converge to the distribution function F_Z for the random variable Z; that is, for each z,

$$\lim_{N \to \infty} G_N(z) = F_Z(z). \tag{3.75}$$

But Z_N is a Gaussian random variable, so

$$G_N(z) = \Phi\left(\frac{z - \mu_N}{\sigma_N}\right), \tag{3.76}$$

where $\mu_N = E\{Z_N\}$, $\sigma_N = \sqrt{\text{Var}\{Z_N\}}$, and Φ is the standard Gaussian distribution function.

Now, recall that

$$\lim_{N\to\infty} Z_N = Z \text{ (mean square)}$$

means that

$$\lim_{N\to\infty} E\{[Z_N - Z]^2\} = 0.$$

It should be intuitively clear (and it can be proven rigorously) that this implies that

$$\lim_{N\to\infty} E\{Z_N\} = E\{Z\}$$

and

$$\lim_{N\to\infty} \text{Var}\{Z_N\} = \text{Var } Z.$$

That is, if $\mu = E\{Z\}$ and $\sigma^2 = \text{Var}\{Z\}$, then

$$\lim_{N\to\infty} \mu_N = \mu$$

and

$$\lim_{N\to\infty} \sigma_N = \sigma.$$

The latter two results imply that for each z,

$$\lim_{N\to\infty} \frac{(z - \mu_N)}{\sigma_N} = \frac{(z - \mu)}{\sigma}.$$

Because Φ is a continuous function, it follows that

$$\lim_{N\to\infty} \Phi\left(\frac{z - \mu_N}{\sigma_N}\right) = \Phi\left(\lim_{N\to\infty}\left(\frac{z - \mu_N}{\sigma_N}\right)\right) = \Phi\left(\frac{z - \mu}{\sigma}\right). \tag{3.77}$$

Finally, we combine (3.76) and (3.77) to obtain

$$\lim_{N\to\infty} G_N(z) = \lim_{N\to\infty} \Phi\left(\frac{z - \mu_N}{\sigma_N}\right) = \Phi\left(\frac{z - \mu}{\sigma}\right).$$

In view of (3.75), it must be that

$$F_Z(z) = \Phi\left(\frac{z - \mu}{\sigma}\right).$$

That is, Z is Gaussian with mean μ and variance σ^2. This concludes the outline of the proof.

The foregoing result is one of the most important properties of Gaussian random processes. Stated concisely, this key property is as follows:

If the input to a linear system is a Gaussian random process, the output is also a Gaussian random process.

This statement is valid for both continuous-time and discrete-time linear systems and random processes. All that is required is that the output process be well defined (i.e., (3.21) or (3.25) must be satisfied). In fact, under the same conditions, the following property holds:

If the input to a linear system is a Gaussian random process, the input and output processes are jointly Gaussian.

The latter property means that for any $t_1, t_2, \ldots, t_n$ and any $m < n$, the collection $X(t_1), \ldots, X(t_m), Y(t_{m+1}), \ldots, Y(t_n)$ of random variables has an n-dimensional Gaussian density function.

In particular, for any t and any s, the joint density function for the random variables $X(t)$ and $Y(s)$ is the bivariate Gaussian density with parameters

$$\mu_1 = E\{X(t)\}, \qquad \mu_2 = E\{Y(s)\},$$
$$\sigma_1^2 = \text{Var}\{X(t)\}, \qquad \sigma_2^2 = \text{Var}\{Y(s)\},$$

and

$$\rho = \frac{\text{Cov}\{X(t), Y(s)\}}{\sigma_1 \sigma_2} = \frac{C_{X,Y}(t,s)}{\sqrt{C_x(t,t) C_Y(s,s)}}.$$

All of these parameters can be evaluated by using the methods of Section 3.3. Of course, the marginal density of $Y(s)$ is the univariate Gaussian density with mean $\mu_Y(s)$ and variance $C_Y(s,s)$. Consequently, the univariate (i.e., one-dimensional) distribution function for $Y(s)$ is given by

$$F_{Y,1}(y; s) = \Phi\left(\frac{y - \mu_Y(s)}{\sqrt{C_Y(s,s)}}\right)$$

for all y. A typical application is given in the next exercise.

Exercise 3–14. In Exercise 3–9, suppose that the input process is Gaussian and that $\tau_0 = 3t_0$. Find the probability that the output at time t exceeds the threshold value $\gamma = 2$.

Solution. Since the input is Gaussian, the output is Gaussian. The mean of the input is zero, so $Y(t)$ has mean $\mu_Y(t) = 0$. According to the solution to Exercise 3–9,

$$\text{Var}\{Y(t)\} = (\sigma_Y)^2 = t_0^2 (3\tau_0)^{-1}[3\tau_0 - t_0].$$

Notice that neither the mean nor the variance of $Y(t)$ depends on t. In fact, the process $Y(t)$ is wide-sense stationary, since the input process is wide-sense stationary and the linear system is time invariant. Since $Y(t)$ is also Gaussian, it is (strictly) stationary. As a result, $P[Y(t) > \gamma]$ does not depend on t. In fact, we can conclude that

$$P[Y(t) > \gamma] = 1 - \Phi(\gamma/\sigma_Y).$$

For $\tau_0 = 3t_0$, $\sigma_Y = 2\sqrt{2}t_0/3$. For $\gamma = 2$,

$$P[Y(t) > \gamma] = P[Y(t) > 2] = 1 - \Phi[3/(\sqrt{2}t_0)].$$

For instance, if $t_0 = \sqrt{2}$, then $P[Y(t) > 2] = 1 - \Phi(1.5) \approx 1 - (.9332) = .0668.$ ■

An important situation that arises in practice, but is not covered by the preceding results, is when the continuous-time input process is *not* a second-order random process. Suppose a continuous-time linear system has an impulse response that satisfies

$$\int_{-\infty}^{\infty} [h(t, \lambda)]^2 \, d\lambda < \infty, \quad \text{for each } t. \tag{3.78}$$

It is known that if the input to this system is thermal noise, the output is a *Gaussian* random process with zero mean and autocorrelation function

$$R_Y(t, s) = \frac{N_0}{2} \int_{-\infty}^{\infty} h(s, \lambda) h(t, \lambda) \, d\lambda, \qquad (3.79)$$

where $N_0/2$ is called the *two-sided power spectral density* of the thermal noise process. In terms of the parameters of Example 2–2, $N_0 = 4kTR$, which is usually expressed in units of watts per Hz (i.e., joules), treating N_0 as a power density, where the power is measured on a one-ohm basis.

Notice that (3.78) is just the necessary and sufficient condition for $R_Y(t, s) < \infty$ for all t and s. This fact follows from (3.79) and (2.17). Notice also that (3.79) can be derived from (3.42) if we set

$$R_X(\tau, \lambda) = \frac{N_0}{2} \delta(\tau - \lambda).$$

To show this, we use the defining property of the Dirac delta function: For any "well-behaved" function g,

$$g(\lambda) = \int_{-\infty}^{\infty} g(\tau)\delta(\tau - \lambda) \, d\tau, \quad \text{for each } \lambda.$$

As pointed out in Chapter 2, a white-noise process (whether it is Gaussian or not) is not a second-order random process. Hence, we must be careful in working with white noise in linear systems. Just as with the delta function in deterministic signal analysis, the autocorrelation function for white noise causes no real difficulties if it appears in the integrand of a mathematical expression.

Because the second moment of white noise is infinite, we have no guarantee that the mean exists. Hence, we do not define the mean of white noise, but we acknowledge that white noise produces a zero-mean output when it is the input of a linear system that satisfies (3.78). Hence, in such a linear system, white noise *behaves* as though it were a *zero-mean* random process. Notice that the autocorrelation function for the white-noise process is a function of the time difference only. As a result, we usually write the autocorrelation function as

$$R_X(\tau) = \frac{N_0}{2} \delta(\tau), \quad -\infty < \tau < \infty.$$

Thus, we can think of white noise as a wide-sense stationary random process, even though it is not a second-order random process.

The foregoing discussion motivates the following definition:

Definition.   *White Gaussian noise is a random process which has the property that when it is the input to a linear system that satisfies (3.78), the output is a zero-mean Gaussian process with autocorrelation function given by (3.79).*

As a practical matter, it is more convenient to think of white Gaussian noise as a random process with the following characteristics:

For engineering applications, a white Gaussian noise process $X(t)$ can be viewed as a stationary Gaussian random process with autocorrelation function given by

$$R_X(\tau) = \frac{N_0}{2}\delta(\tau).$$

To illustrate the kind of manipulations that are valid in dealing with white noise, we derive the crosscorrelation function for a continuous-time linear system that has a white-noise input. First, in (3.40), let

$$R_X(\tau, s) = \frac{N_0}{2}\delta(\tau - s)$$

to obtain

$$R_{Y,X}(t, s) = \frac{N_0}{2}\int_{-\infty}^{\infty} h(t, \tau)\delta(\tau - s)\, d\tau.$$

Next, we use the defining property of the delta function to conclude that

$$R_{Y,X}(t, s) = \frac{N_0}{2}h(t, s).$$

This result is the basis for an interesting approach to the experimental determination of the impulse response of a linear system (i.e., system identification), a topic that is discussed in Section 3.6.

The preceding results all simplify somewhat for time-invariant systems. In particular, (3.78) and (3.79) become

$$\int_{-\infty}^{\infty} [h(t)]^2\, dt < \infty,$$

and by letting $u = \lambda - s$, we obtain

$$R_Y(t, s) = \frac{N_0}{2}\int_{-\infty}^{\infty} h(-u)h(t - s - u)\, du$$

$$= \frac{N_0}{2}\int_{-\infty}^{\infty} \tilde{h}(u)h(t - s - u)\, du,$$

where $\tilde{h}(t) = h(-t)$. Because $Y(t)$ has zero mean, it follows that $Y(t)$ is wide-sense stationary and has autocorrelation function

$$R_Y(\tau) = \frac{N_0}{2}\int_{-\infty}^{\infty} h(-u)h(\tau - u)\, du$$

$$= \frac{N_0}{2}\int_{-\infty}^{\infty} \tilde{h}(u)h(\tau - u)\, du. \qquad (3.80)$$

In other words, $R_Y(\tau) = (N_0/2)f(\tau)$, where $f = \tilde{h} * h$, a relationship that can be derived directly from (3.54). So if the input to a time-invariant linear system with impulse response h is white Gaussian noise, then the output $Y(t)$ is a zero-mean Gaussian random process with autocorrelation function

$$R_Y(\tau) = \frac{N_0}{2} f(\tau),$$

where $f = \tilde{h} * h$. The actual calculation of $R_Y(\tau)$ is usually easier if we employ the fact that $R_Y(\tau) = R_Y(-\tau)$ to observe from (3.80) that

$$R_Y(\tau) = \frac{N_0}{2} \int_{-\infty}^{\infty} h(-u)h(-\tau - u)\, du = \frac{N_0}{2} \int_{-\infty}^{\infty} h(t)h(t - \tau)\, dt. \qquad (3.81)$$

Exercise 3–15. Suppose that white Gaussian noise with spectral density $N_0/2$ is the input to a time-invariant linear filter with impulse response $2\exp(-4t)$. The output process, denoted by $Y(t)$, is sampled at times $t = 0$ and $t = 1$. The samples are added to give the random variable $W = Y(0) + Y(1)$. Find the probability that the random variable W is greater than 5.

Solution. First, observe that $Y(t)$ is a Gaussian, wide-sense stationary random process, because the input is white Gaussian noise. Therefore, the distribution function for W is $F_W(u) = \Phi[(u - \mu)/\sigma]$, where μ is the mean, and σ^2 is the variance, of W. The probability that we are asked to determine is therefore given by $P(W > 5) = 1 - \Phi[(5 - \mu)/\sigma]$. Next notice that $\mu = 0$, because $Y(t)$ has zero mean. It follows from this that σ^2 is just the second moment of W, which is given by $E\{W^2\} = 2R_Y(0) + 2R_Y(1)$. The solution to Exercise 3–7 ($\beta = 2$ and $\alpha = 4$) shows that $R_Y(0) = N_0/4$ and $R_Y(1) = N_0 e^{-4}/4$; therefore, $\sigma^2 = E\{W^2\} = N_0(1 + e^{-4})/2$. Combining all of these facts, we see that $P(W > 5) = 1 - \Phi[5\sqrt{2}/\{N_0(1 + \exp(-4))\}]$. ∎

The next example further illustrates the analysis of white noise in linear systems, and it also introduces a new random process.

Example 3–5 The Wiener–Levy Process

Suppose the input to the system described in the solution to Exercise 3–10 is white Gaussian noise. That is, the random process $Y(t), t \geq 0$, is defined by

$$Y(t) = \int_0^t X(\tau)\, d\tau,$$

where $X(t)$ is a white Gaussian noise process. In the alternative approach given as part of the solution to that exercise, we found that

$$R_Y(t, s) = \int_0^t \int_0^s R_X(\tau - \lambda)\, d\lambda\, d\tau.$$

Thus, if $X(t)$ is white noise with spectral density $N_0/2$, then

$$R_Y(t, s) = \frac{N_0}{2} \int_0^t \int_0^s \delta(\tau - \lambda)\, d\lambda\, d\tau.$$

If $0 \le t \le s$, then, for each τ in the range $0 \le \tau \le t$, there is a value of λ in the range $0 \le \lambda \le s$ for which $\lambda = \tau$. It follows that, for $t < s$,

$$\int_0^s \delta(\tau - \lambda)\, d\lambda = 1,$$

and thus,

$$R_Y(t, s) = \frac{N_0}{2} \int_0^t 1 \, d\tau = \frac{N_0 t}{2}.$$

On the other hand, for $0 \le s < t$, there are values of τ in $[0, t]$ for which there is no λ in $[0, s]$ such that $\lambda = \tau$ [namely, those τ's in the interval (s, t)]. Thus, for $t > s$, we consider the reverse order of integration to obtain

$$R_Y(t, s) = \frac{N_0}{2} \int_0^s \int_0^t \delta(\tau - \lambda)\, d\tau\, d\lambda,$$

to which the argument in the preceding paragraph applies. (Just reverse the roles of t and s.) Hence, for $t > s$,

$$\int_0^t \delta(\tau - \lambda)\, d\tau = 1$$

and

$$R_Y(t, s) = \frac{N_0}{2} \int_0^s 1 \, d\lambda = \frac{N_0 s}{2}.$$

The conclusion is that $Y(t)$, $t \ge 0$, is a zero-mean Gaussian random process with autocorrelation function

$$R_Y(t, s) = \begin{cases} N_0 t/2, & t \le s, \\ N_0 s/2, & t > s. \end{cases}$$

This autocorrelation function can be written in the more compact form

$$R_Y(t, s) = \frac{N_0}{2} \min(t, s)$$

for all $t \ge 0$ and $s \ge 0$. Notice that $\min(t, s)$ does not depend on $t - s$ only. Notice also that

$$\text{Var}\{Y(t)\} = R_Y(t, t) = \frac{N_0 t}{2},$$

so that the variance of $Y(t)$ depends on t. The random process $Y(t)$, $t \ge 0$, obtained in this manner is the *Wiener–Lévy process* (often called the Brownian motion process). It is a *nonstationary*, second-order Gaussian process.   ∎

3.5   DETERMINATION OF CORRELATION FUNCTIONS

In all of the preceding examples and exercises that involve either autocorrelation functions or crosscorrelation functions, we have assumed that the autocorrelation function of at least one of the random processes (e.g., the input process) is known. If $X(t)$ is the random process with known autocorrelation function, the problem is to find the autocorrelation function for a related random process $Y(t)$ or to find the crosscorrelation function for the two random processes $X(t)$ and $Y(t)$. Also, for problems in which $X(t)$ and $Y(t)$ are the system input and output, respectively, we have assumed that the impulse response of the system is known.

Actually, these assumptions are quite often valid in practice. For instance, it is frequently true in a satellite communication system that the input at some point in the system is white noise (described in Section 2.6), which implies that the autocorrelation function of this process is a delta function. The impulse responses for the linear filters in such a system are known quite accurately in practice, so autocorrelation functions and crosscorrelation functions can be determined for the processes at the outputs of these filters.

However, for certain engineering design and analysis problems, such correlation functions may be unknown and not directly computable, by using only the methods of this chapter, from the information that is known. If this is the case, the first step in the solution to the problem is to determine the correlation functions for the processes of interest. Usually, this is accomplished in one of two ways: If enough information is known about a random process, its autocorrelation function can be determined analytically; otherwise, the autocorrelation function is determined experimentally. Each of these approaches is illustrated in this section. First we give an example in which the analytical approach is used, and then we describe a general technique for the experimental determination of correlation functions for wide-sense stationary processes.

3.5.1   The Autocorrelation Function for a Poisson Process

Consider for the moment the counting process described in Example 2–3. Suppose we are modeling a particular random process as a counting process for which we require the number of events that occur in one time interval to be independent of the number of events that occur in all other nonoverlapping time intervals. For instance, in most situations, the number of phone calls received at a switchboard on Wednesday can be modeled as a random variable that is independent of the number of phone calls received on Tuesday. Secondly, we require that the probability distribution for the number of events occurring in the time interval $(t, t + \tau)$ be the same for all $t \geq 0$. That is, the probability $p_n(\tau)$ of exactly n events occurring in a time interval of length τ is the same, regardless of the time at which the interval starts.

Mathematically, we describe such a process as follows: Let $t_0 > 0$ be arbitrary, and let $t_1, t_2, \ldots, t_{2K}$ be such that $0 \leq t_i \leq t_j$ whenever $1 \leq i \leq j \leq 2K$. Consider the random vector $\mathbf{Z} = (Z_1, Z_2, \ldots, Z_K)$ for which $Z_k = X(t_{2k} + t_0) - X(t_{2k-1} + t_0)$ for $1 \leq k \leq K$. If $X(t)$ is the counting process just described, then the components $Z_1, Z_2, \ldots, Z_K$ are mutually independent, and the distribution function for $\mathbf{Z}$ does not depend on t_0.

For further consideration of this random process, it is helpful to rely on some standard terminology. First, for $s < t$, the quantity $X(t) - X(s)$ is called an *increment* of the random process $X(t)$. If, for each positive integer K and each choice of $t_0, t_1, t_2, \ldots, t_{2K}$, the distribution function for the random vector $\mathbf{Z}$ does not depend on t_0, the process $X(t)$ is said to have *stationary increments*. If, for each choice of K and $t_0, t_1, t_2, \ldots, t_{2K}$, the random variables $Z_1, Z_2, \ldots, Z_K$ are mutually independent, the process $X(t)$ is said to have *independent increments*. Notice that if $X(t)$ has independent increments, then, for $0 < s < t$, the random variables $Z_1 = X(s) - X(0)$ and $Z_2 = X(t) - X(s)$ are independent, which implies that they are also *uncorrelated*. Thus,

$$E\{Z_1 Z_2\} = E\{[X(s) - X(0)][X(t) - X(s)]\}$$
$$= E\{X(s) - X(0)\}E\{X(t) - X(s)\}, \tag{3.82}$$

a fact that will be useful later. A random process with independent increments also has *uncorrelated increments*, but, in general, the reverse implication is not true.

Recall that in Example 2–3 we required that all jumps or steps in the process must be of size one; that is, no more than one event can take place at a time. Stated more precisely, this requirement is that the probability of two or more events occurring at exactly the same time is zero. Whenever we use the term *counting process* for a random process $X(t)$, we assume that $X(0) = 0$, $X(t) \le X(t + \tau)$ for all $t \ge 0$ and all $\tau \ge 0$, and at most one event can take place at a time. These assumptions are made in Example 2–3.

It is known that for any counting process that has stationary, independent increments, there is a constant $\nu > 0$ for which the probability of exactly n events occurring ·in a time interval of length τ is given by

$$p_n(\tau) = (\nu\tau)^n e^{-\nu\tau}/n!$$

for all $\tau \ge 0$ and all nonnegative integers n. In other words, for each $t \ge 0$, $X(t)$ has the Poisson distribution with parameter $\lambda = \nu t$. The process $X(t)$ is called a *Poisson process*.

Our real interest at the moment is in the autocorrelation function for such a process. First, recall that a random variable that has the Poisson distribution with parameter λ has mean λ and variance λ. Thus, if $X(t)$ is a Poisson process, then

$$E\{X(t)\} = \nu t$$

and

$$\text{Var}\{X(t)\} = \nu t$$

for any $t > 0$. These relationships imply that

$$E\{[X(t)]^2\} = \nu t + (\nu t)^2 = \nu t(1 + \nu t). \tag{3.83}$$

Next, we assume that $0 < t_1 < t_2$ and evaluate

$$R_X(t_1, t_2) = E\{X(t_1)X(t_2)\}$$
$$= E\{X(t_1)[X(t_2) - X(t_1) + X(t_1)]\}$$
$$= E\{X(t_1)[X(t_2) - X(t_1)]\} + E\{[X(t_1)]^2\}. \tag{3.84}$$

Now, from (3.83), we have

$$E\{[X(t_1)]^2\} = \nu t_1(1 + \nu t_1),$$

and because $X(0) = 0$,

$$E\{X(t_1)[X(t_2) - X(t_1)]\} = E\{[X(t_1) - X(0)][X(t_2) - X(t_1)]\}.$$

But the process $X(t)$ has independent increments, so, from (3.82), we have

$$E\{X(t_1)[X(t_2) - X(t_1)]\} = E\{X(t_1)\}E\{X(t_2) - X(t_1)\} = \nu t_1(\nu t_2 - \nu t_1),$$

where we have used the fact that $E\{X(t_i)\} = \nu t_i$. Consequently, (3.84) implies that

$$R_X(t_1, t_2) = \nu t_1(\nu t_2 - \nu t_1) + \nu t_1(1 + \nu t_1) = \nu t_1(1 + \nu t_2).$$

Similarly (just interchange the roles of t_1 and t_2), we find that for $0 < t_2 < t_1$,

$$R_X(t_1, t_2) = \nu t_2(1 + \nu t_1).$$

These two cases can be combined to give

$$R_X(t_1, t_2) = \nu \min(t_1, t_2)[1 + \nu \max(t_1, t_2)], \tag{3.85}$$

which is valid for all $t_1 \geq 0$ and $t_2 \geq 0$. For a process with nonzero mean, such as the Poisson process, it is often easier to work with the autocovariance function $C_X(t_1, t_2)$. From (3.85), it is easy to show that

$$C_X(t_1, t_2) = \nu \min(t_1, t_2).$$

This relationship follows from $E\{X(t_i)\} = \nu t_i$,

$$R_X(t_1, t_2) = C_X(t_1, t_2) + E\{X(t_1)\}E\{X(t_2)\},$$

and the fact that

$$\min(t_1, t_2) \max(t_1, t_2) = t_1 t_2.$$

Notice that the Poisson process has the same autocovariance function as the Wiener–Lévy process.

The main point is that the autocorrelation and autocovariance functions are obtained *analytically* on the basis of some physical properties of the process. In fact, many of the key properties that we used (e.g., stationarity and independence) are *qualitative* rather than *quantitative*. This is in contrast to the experimental approach described next, wherein we seek to obtain directly some quantitative information about the correlation function (e.g., the value of $R_X(2, 0)$). The analytical approach, whenever it can be employed, typically gives solutions to a wider class of problems. In the foregoing analysis, we obtained the autocorrelation function for the class of all Poisson processes. Furthermore, along the way, we established some results that are valid for any independent increment process [e.g., (3.82)]. In fact, a review of the derivation will reveal that we really used only the property that the increments are *uncorrelated*, a property which is considerably weaker than the independence of the increments.

It is not correct to infer from the preceding discussion that experimental methods are without merit or that they are unnecessary for engineering problems involving random processes. Experimental results are extremely useful in many such problems. Even in the example of the Poisson process, notice that we have not yet completely solved the problem, because our solution is in terms of the parameter ν. In practice, experimentation may be needed to determine the correct value of ν for a given application. However, it is generally more efficient to estimate only a single parameter than to attempt to estimate $R_X(t_1, t_2)$ for each different value of t_1 and t_2. A simple estimate for ν is $\hat{\nu} = x(T_0)/T_0$, which is based on observing the sample function $x(t)$, $0 \leq t \leq T_0$, from the process $X(t)$. Notice that this is just the average number of jumps per unit time in the sample function $x(t)$.

3.5.2   Experimental Determination of Means and Correlation Functions for Ergodic Random Processes

In this section, we restrict attention to random processes that are *ergodic*, which, for our purposes, means that almost all sample functions for the random process are typical of the entire process. For such processes, certain time averages converge to probabilistic averages as the time interval for the averaging becomes large. Actually, in what follows, we really require only a considerably weaker property, which is often referred to as *weak ergodicity*. A stationary Gaussian random process, for example, is weakly ergodic if its autocovariance function is absolutely integrable. In many situations, even if we do not know the autocovariance function precisely, we may know that it is absolutely integrable (e.g., it may be enough to have a bound on the autocovariance function).

Suppose we can devise an experiment that allows us to observe a sample function of an ergodic, wide-sense stationary, random process $X(t)$ for an arbitrarily long period of time. On the basis of this observation, we can estimate the autocorrelation function of $X(t)$ with any specified degree of accuracy. More generally, given sample functions from a jointly ergodic, jointly wide-sense stationary pair of processes $X(t)$ and $Y(t)$, we can estimate the crosscorrelation function of the two processes. Since the latter problem includes the former as a special case (simply let $X(t) = Y(t)$), we examine the estimation of the crosscorrelation function.

Consider the system shown in Figure 3–8. The output of the integrator is given by

$$Z(\tau) = T_0^{-1} \int_\tau^{T_0+\tau} Y(t)X(t-\tau)\,dt = T_0^{-1} \int_0^{T_0} Y(t+\tau)X(t)\,dt, \qquad (3.86)$$

where τ is arbitrary, but held constant for the moment. For sufficiently large values of T_0, $Z(\tau)$ is a good approximation to $R_{Y,X}(\tau)$ for most of the jointly wide-sense stationary random processes that are encountered in engineering (i.e., those which have the joint ergodicity property). Notice, for instance, that

$$E\{Z(\tau)\} = T_0^{-1} \int_0^{T_0} E\{Y(t+\tau)X(t)\}\,dt$$
$$= R_{Y,X}(\tau)$$

for any $T_0 > 0$.

A commonly used measure of accuracy involved in estimating $R_{Y,X}(\tau)$ by $Z(\tau)$ is the mean-squared error $E\{[Z(\tau) - R_{Y,X}(\tau)]^2\}$. For ergodic processes, the mean-squared error will converge to zero as $T_0 \to \infty$. In particular, the mean-squared error

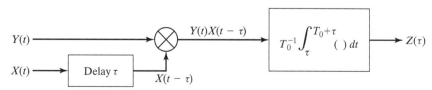

FIGURE 3–8   Estimating a crosscorrelation function.

will converge to zero for any stationary jointly Gaussian random processes $X(t)$ and $Y(t)$ for which

$$\int_{-\infty}^{\infty} |C_X(\tau)| \, d\tau < \infty \tag{3.87}$$

and

$$\int_{-\infty}^{\infty} |C_Y(\tau)| \, d\tau < \infty. \tag{3.88}$$

A related, but somewhat simpler, problem is to find the *mean* of a wide-sense stationary random process $X(t)$. In this case, we simply use the integrator of Figure 3–8 with $X(t)$ alone as its input. The output is the random variable

$$W = T_0^{-1} \int_0^{T_0} X(t) \, dt,$$

which has the property

$$E\{W\} = E\{X(t)\} = \mu_X.$$

The mean-squared error for this estimate depends on the value of T_0 and is given by

$$\begin{aligned}
\varepsilon(T_0) &= E\{[W - \mu_X]^2\} \\
&= E\left\{\left[T_0^{-1} \int_0^{T_0} X(t) \, dt - \mu_X\right]^2\right\} \\
&= E\left\{\left[T_0^{-1} \int_0^{T_0} (X(t) - \mu_X) \, dt\right]^2\right\} \\
&= E\left\{\left[T_0^{-1} \int_0^{T_0} Y(t) \, dt\right]^2\right\},
\end{aligned}$$

where $Y(t) = X(t) - \mu_X$. The problem of finding the second moment of the integral of a random process is a problem that we have already solved. In the solution to Exercise 3–10, we found that

$$\begin{aligned}
E\left\{\left[\int_0^{T_0} Y(t) \, dt\right]^2\right\} &= \int_0^{T_0} \int_0^{T_0} R_Y(t - s) \, dt \, ds \\
&= 2 \int_0^{T_0} (T_0 - \lambda) R_Y(\lambda) \, d\lambda.
\end{aligned}$$

[See (3.56) for instance.] Next, we notice that $R_Y(\lambda) = C_X(\lambda)$ and that, for $0 \le \lambda \le T_0$,

$$0 \le (T_0 - \lambda) \le T_0.$$

Since $E\{[W - \mu_X]^2\} \ge 0$, we have

$$0 \le \varepsilon(T_0) = 2T_0^{-2} \int_0^{T_0} (T_0 - \lambda) C_X(\lambda) \, d\lambda$$

$$\leq 2T_0^{-2} \int_0^{T_0} (T_0 - \lambda)|C_X(\lambda)| \, d\lambda$$

$$\leq 2T_0^{-1} \int_0^{T_0} |C_X(\lambda)| \, d\lambda. \tag{3.89}$$

If the process $X(t)$ satisfies (3.87), then

$$\lim_{T_0 \to \infty} \varepsilon(T_0) = \left(\lim_{T_0 \to \infty} T_0^{-1} \right) \left(\lim_{T_0 \to \infty} 2 \int_0^{T_0} |C_X(\lambda)| \, d\lambda \right)$$

$$= \left(\lim_{T_0 \to \infty} T_0^{-1} \right) \int_{-\infty}^{\infty} |C_X(\lambda)| \, d\lambda = 0.$$

Hence, (3.87) and (3.89) imply that the mean-squared error converges to zero as $T_0 \to \infty$. Furthermore, for any finite T_0, (3.89) provides the following upper bound on the mean-squared error:

$$\varepsilon(T_0) \leq 2T_0^{-1} \int_0^{T_0} |C_X(\lambda)| \, d\lambda$$

$$\leq T_0^{-1} \int_{-\infty}^{\infty} |C_X(\lambda)| \, d\lambda = T_0^{-1} \Gamma_X.$$

In the rightmost term,

$$\Gamma_X = \int_{-\infty}^{\infty} |C_X(\lambda)| \, d\lambda < \infty.$$

Hence, if $C_X(\tau)$ is known, then Γ_X can be determined, and a prespecified mean-square error ε_{ms} can be guaranteed by selecting T_0 to satisfy $T_0 \geq \Gamma_X / \varepsilon_{\text{ms}}$.

3.6   APPLICATIONS

The material presented in this chapter is applicable to such fields as communications, control, radar, signal processing, and system design. Unfortunately, for many of the key applications that arise in these disciplines, further, more specialized training in the particular field is required in order to understand and appreciate the applications. The next three examples provide a limited insight based on the results obtained thus far.

Example 3–6   System Identification

Suppose that the impulse response of a continuous-time, time-invariant linear system is unknown. Theoretically, we could use a delta function as an input and observe the response. Letting $x(\tau) = \delta(\tau)$ in (3.1), we find that $y(t) = h(t)$. The problem is that delta functions are impossible to generate in the laboratory (as are ideal step functions and related signals), and even if one could be generated, its effect on the system would be disastrous. An alternative approach is to let the input to the system be white Gaussian noise (see Section 3.4) or at least noise whose spectrum is flat over a band wider than the bandwidth of the system we are trying to identify; usually, a rough estimate of the system's bandwidth will suffice. White Gaussian noise in the form of thermal noise is relatively easy to generate: Any resistive component generates thermal noise as a result of the random motion of its conduction electrons. (See Example 2–2.)

If the input to the system is white noise, the autocorrelation function (not the signal itself) is a delta function. Thus, the same effect as that obtained by a delta-function input can be produced without actually generating a delta-function signal. To determine the impulse response h in this manner, we need to know only the crosscorrelation function between the input and output. This fact follows from (3.48) with $R_X(\tau) = \delta(\tau)$, which implies that $R_{Y,X}(\tau) = h(\tau)$. But $R_{Y,X}(\tau)$ can be determined as described in Section 3.5.2. Thus, $h(\tau)$ can be determined by the system shown in Figure 3–8 if $X(t)$ is the input to the linear time-invariant system and $Y(t)$ is the output.   ∎

Example 3–7   Prediction of Gaussian Processes

Consider the problem of predicting the value of a discrete-time Gaussian random process $X(t)$ at some *future* time $t = k$, based on the observation of the value of $X(t)$ at the *present* time $t = m$. (Clearly, $m < k$.) Suppose we wish to require the prediction $\hat{X}(k)$ to have the minimum possible mean-squared error

$$\varepsilon_{\mathrm{ms}} = E\{[X(k) - \hat{X}(k)]^2\}$$

for any prediction that is based only upon the observation of $X(m)$. The optimal predictor turns out to be the conditional expected value of $X(k)$ given $X(m)$; that is,

$$\hat{X}(k) = E\{X(k)|X(m)\}.$$

This is not surprising from an intuitive point of view, since we are simply predicting that $X(k)$ will take on its "average" value, and the "average" should be based on all of the information that is available [namely, the value of $X(m)$]. Formally, this "average" is the conditional expectation given the value of $X(m)$.

The conditional expectation is actually the optimal predictor even if the random process is not Gaussian. However, when the process is Gaussian, the prediction $\hat{X}(k)$ has a very simple form. Recall that if Y_1 and Y_2 are jointly Gaussian with means μ_1 and μ_2, variances σ_1^2 and σ_2^2, and correlation coefficient ρ, then

$$E\{Y_1|Y_2\} = \mu_1 + \rho\sigma_1\sigma_2^{-1}(Y_2 - \mu_2).$$

Since the random process is Gaussian, $X(m)$ and $X(k)$ are jointly Gaussian, so that

$$\begin{aligned}
\hat{X}(k) &= E\{X(k)|X(m)\} \\
&= \mu_X(k) + \rho_X(k,m)\sigma_X(k)[\sigma_X(m)]^{-1}[X(m) - \mu_X(m)], \quad (3.90)
\end{aligned}$$

where

$$\rho_X(k,m) = [\sigma_X(k)\sigma_X(m)]^{-1}C_X(k,m)].$$

That is, given that we observe $X(m) = x$, the minimum mean-squared-error (MMSE) prediction of $X(k)$ is

$$\begin{aligned}
\hat{X}(k) &= E\{X(k)|X(m) = x\} \\
&= \mu_X(k) + [\mathrm{Var}\{X(m)\}]^{-1}C_X(k,m)[x - \mu_X(m)] \\
&= [\mathrm{Var}\{X(m)\}]^{-1}C_X(k,m)x + \{\mu_X(k) - \mu_X(m)[\mathrm{Var}\{X(m)\}]^{-1}C_X(k,m)\}.
\end{aligned}$$
$$(3.91)$$

If the random process is Gaussian *and stationary*, the MMSE prediction based on the observation $X(m) = x$ is given by the simpler expression

$$\begin{aligned}
\hat{X}(k) &= \mu_X + [\sigma_X^2]^{-1}C_X(k - m)[x - \mu_X] \\
&= \mu_X\{1 - \sigma_X^{-2}C_X(k - m)\} + \sigma_X^{-2}C_X(k - m)x.
\end{aligned}$$

Finally, if the stationary Gaussian random process $X(t)$ has *zero mean*, then

$$\hat{X}(k) = \sigma_X^{-2} R_X(k - m)x.$$

Even if the random process is not Gaussian, the predictor given by (3.90) is the MMSE *linear* predictor. (Strictly speaking, what we mean is *affine*, which is "linear plus a constant," but the convention is to call this a linear predictor.) The precise statement of this fact is as follows: Of all predictors of the form

$$\hat{X}(k) = aX(m) + b,$$

where a and b are constants, the one that minimizes the mean-squared error

$$\varepsilon_{\mathrm{ms}} = E\{[X(k) - \hat{X}(k)]^2\} = E\{[X(k) - aX(m) - b]^2\}$$

is obtained by setting $a = C_X(k, m)/\mathrm{Var}\{X(m)\}$ and

$$b = \mu_X(k) - \mu_X(m)\left\{\frac{C_X(k, m)}{\mathrm{Var}\{X(m)\}}\right\} = \mu_X(k) - a\mu_X(m),$$

as in (3.91).   ∎

Example 3–8   Digital Communication System

Consider a baseband communication system in which binary digits are transmitted as rectangular pulses of duration T. A binary zero is transmitted as a positive rectangular pulse $s_0(t) = Ap_T(t)$, and a binary one is transmitted as a negative rectangular pulse $s_1(t) = -Ap_T(t)$, where A is a deterministic, positive constant that represents the amplitude of the signal and

$$p_T(t) = \begin{cases} 1, & 0 \le t < T, \\ 0, & \text{otherwise.} \end{cases}$$

These pulses are transmitted over an additive white Gaussian noise channel, and the signal that is received is

$$Y_i(t) = s_i(t) + X(t),$$

where $i = 0$ or 1, depending on whether the binary digit that is transmitted is 0 or 1, and $X(t)$ is a white Gaussian noise process with spectral density $N_0/2$.

The first receiver consists of a linear time-invariant filter followed by a sampler. The filter is a simple RC filter with $RC = T/2$. Its impulse response is

$$h(t) = (RC)^{-1}e^{-t/RC}u(t),$$

where

$$u(t) = \begin{cases} 1, & t > 0, \\ 0, & t \le 0, \end{cases}$$

is the unit step function. The output $V_i(t)$ of the RC filter is sampled at times $t_1 = T/2$, $t_2 = 3T/4$, and $t_3 = T$. The sample values are then summed to give

$$Z_i = \sum_{k=1}^{3} V_i(t_k).$$

The receiver decides that a 0 was sent if the sum of the sample values exceeds zero, and it decides that a 1 was sent if the sum is less than or equal to zero. A block diagram of the receiver is shown in Figure 3–9.

An alternative receiver for the same transmitted signals $s_0(t)$ and $s_1(t)$ and the same additive white Gaussian noise channel is shown in Figure 3–10. This receiver is optimal for

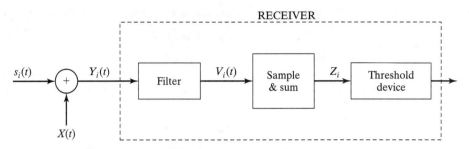

FIGURE 3–9   A communication system with a suboptimal receiver.

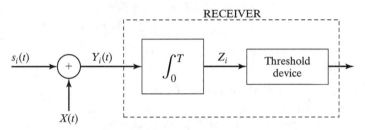

FIGURE 3–10   An optimal receiver.

the given signals and channel in the sense that it gives the minimum possible error probability of any communication receiver.

To clarify what is meant by *minimum possible error probability*, we let $P_{e,i}$ denote the probability of error for a given receiver when the symbol i is sent (i is either 0 or 1). That is, $P_{e,0}$ is the probability that the receiver decides that a 1 was sent when in fact a 0 was sent, and $P_{e,1}$ is the probability that the receiver decides that a 0 was sent when a 1 was actually sent. We then define

$$P_{e,m} = \max\{P_{e,0}, P_{e,1}\}$$

and

$$\bar{P}_e = \frac{P_{e,0} + P_{e,1}}{2}.$$

The receiver of Figure 3–10 gives the smallest possible value of $P_{e,m}$ that can be obtained with any receiver. It also gives the smallest possible value of $\bar{P}_e$.

To illustrate the general analytical methods applicable to this type of problem, and, in particular, to demonstrate the role played by the autocorrelation functions in such methods, we derive expressions for $P_{e,0}$ and $P_{e,1}$ for the receivers of Figures 3–9 and 3–10. This derivation requires the evaluation of

$$P_{e,0} = P(Z_0 \leq 0)$$

and

$$P_{e,1} = P(Z_1 > 0)$$

for each of the two receivers.

For the suboptimal receiver of Figure 3–9, Z_i is a Gaussian random variable with mean

$$\mu_i = E\left\{\sum_{k=1}^{3} V_i(t_k)\right\} = \sum_{k=1}^{3} v_i(t_k)$$

and variance

$$\sigma_i^2 = \text{Var}\left\{\sum_{k=1}^{3} V_i(t_k)\right\}$$

$$= E\left\{\left[\sum_{k=1}^{3} \{V_i(t_k) - v_i(t_k)\}\right]^2\right\}, \tag{3.92}$$

where

$$v_i(t_k) = E\{V_i(t_k)\}.$$

Now, $V_i(t)$ consists of the sum of a signal component

$$\hat{s}_i(t) = \int_{-\infty}^{\infty} h(\tau)s_i(t - \tau)\, d\tau$$

and a noise component

$$\hat{X}(t) = \int_{-\infty}^{\infty} h(\tau)X(t - \tau)\, d\tau.$$

That is, $V_i(t) = \hat{s}_i(t) + \hat{X}(t)$, where $\hat{s}_i = s_i * h$ is the output of the filter due to the signal alone and $\hat{X}(t)$ is the output of the filter due to noise alone. Since the noise $X(t)$ is white, the output process $\hat{X}(t)$ has zero mean and an autocorrelation function R_X given by

$$R_X(\tau) = \frac{N_0}{2} \int_{-\infty}^{\infty} h(u)h(u + \tau)\, du.$$

Because $\hat{X}(t)$ has zero mean,

$$v_i(t) = E\{V_i(t)\} = E\{\hat{s}_i(t) + \hat{X}(t)\}$$

$$= \hat{s}_i(t) + E\{\hat{X}(t)\} = \hat{s}_i(t).$$

Thus, the mean of $V_i(t)$ is just its signal component. Consequently, from (3.92) we see that

$$\sigma_i^2 = E\left\{\left[\sum_{k=1}^{3} \hat{X}(t_k)\right]^2\right\}$$

$$= E\left\{\sum_{k=1}^{3}\sum_{j=1}^{3} \hat{X}(t_k)\hat{X}(t_j)\right\}$$

$$= \sum_{k=1}^{3}\sum_{j=1}^{3} R_{\hat{X}}(t_k - t_j).$$

For given values of t_1, t_2, and t_3,

$$|t_1 - t_2| = |t_2 - t_3| = \frac{T}{4}$$

and

$$|t_1 - t_3| = \frac{T}{2}.$$

Thus,

$$\sigma_i^2 = 3R_{\hat{X}}(0) + 4R_{\hat{X}}(T/4) + 2R_{\hat{X}}(T/2). \tag{3.93}$$

Notice, in particular, that σ_i^2 does not depend on i (i.e., $\sigma_1^2 = \sigma_0^2$). For simplicity, we drop the subscripts and denote the common value of σ_1^2 and σ_0^2 by σ^2.

Next, we must find $v_i(t)$ and $R_{\hat{X}}(\tau)$. The first step is to evaluate

$$v_i(t) = \hat{s}_i(t) = \int_{-\infty}^{\infty} h(\tau)s_i(t - \tau)\, d\tau$$

$$= (-1)^i \frac{A}{RC} \int_0^{\infty} e^{-\tau/RC} p_T(t - \tau)\, d\tau.$$

Since $p_T(t - \tau) = 0$ for $t < 0$, it follows that $v_i(t) = 0$ for $t < 0$. For $0 \le t < T$,

$$v_i(t) = (-1)^i \frac{A}{RC} \int_0^t e^{-\tau/RC}\, d\tau$$

$$= (-i)^i A[1 - \exp(-t/RC)],$$

and for $t \ge T$,

$$v_i(t) = (-1)^i \frac{A}{RC} \int_{t-T}^t \exp(-\tau/RC)\, d\tau$$

$$= (-1)^i A[\exp(T/RC) - 1]\exp(-t/RC)].$$

Thus, the mean of Z_i is

$$\mu_i = (-1)^i A[3 - e^{-T/2RC} - e^{-3T/4RC} - e^{-T/RC}]$$

$$= (-1)^i A[3 - e^{-1} - e^{-3/2} - e^{-2}],$$

where, in the last step, we have used the fact that $RC = T/2$. The autocorrelation function for $\hat{X}(t)$ is given by

$$R_{\hat{X}}(\tau) = \frac{N_0}{2}(RC)^{-2} \int_{-\infty}^{\infty} e^{-u/RC} e^{-(u+\tau)/RC}\, du$$

for $\tau \ge 0$. For $\tau < 0$, we can use the fact that $R_{\hat{X}}(\tau) = R_{\hat{X}}(-\tau)$.

Evaluating the integral in the expression for the autocorrelation function, we find that for $\tau \ge 0$,

$$R_{\hat{X}}(\tau) = \frac{N_0}{4RC} e^{-\tau/RC} = \frac{N_0}{2T} e^{-2\tau/T}.$$

Therefore, since $R_{\hat{X}}(\tau) = R_{\hat{X}}(-\tau)$, it follows that

$$R_{\hat{X}}(\tau) = \frac{N_0}{2T} \exp(-2|\tau|/T),$$

for all τ. From (3.93), we see that the variance of Z_i, which does not depend on i, is given by

$$\sigma^2 = \frac{N_0}{2T}[3 + 4e^{-1/2} + 2e^{-1}].$$

The error probability $P_{e,0}$ is given by

$$P_{e,0} = P(Z_0 \leq 0) = \Phi\left(\frac{0 - \mu_0}{\sigma}\right) = 1 - \Phi\left(\frac{\mu_0}{\sigma}\right),$$

and the error probability $P_{e,1}$ is given by

$$P_{e,1} = P(Z_1 > 0) = 1 - P(Z_1 \leq 0) = 1 - \Phi\left(\frac{0 - \mu_1}{\sigma}\right) = 1 - \Phi\left(\frac{-\mu_1}{\sigma}\right).$$

Evaluating the expressions for μ_i and σ, we find that

$$\mu_i \approx (2.2737)(-1)^i A$$

and

$$\sigma \approx 2.4823\sqrt{\frac{N_0}{2T}}.$$

Hence,

$$\frac{\mu_i}{\sigma} \approx (-1)^i 0.9159\sqrt{\frac{2A^2T}{N_0}}$$

$$= (-1)^i 0.9159\sqrt{\frac{2\mathcal{E}_b}{N_0}},$$

where $\mathcal{E}_b = A^2T$ is the energy per bit; that is,

$$\mathcal{E}_b = \int_{-\infty}^{\infty} [s_i(t)]^2 \, dt = A^2T.$$

Next, we notice that

$$P_{e,0} = P_{e,1} = 1 - \Phi\left(\frac{|\mu_i|}{\sigma}\right)$$

$$\approx 1 - \Phi\left[(0.9159)\sqrt{\frac{2\mathcal{E}_b}{N_0}}\right].$$

If $\mathcal{E}_b/N_0 = 5$, for instance, then

$$P_{e,0} = P_{e,1} \approx 1 - \Phi[0.9159\sqrt{10}]$$

$$\approx 1 - \Phi(2.896)$$

$$\approx 0.00189 = 1.89 \times 10^{-3}.$$

The optimal receiver is actually much simpler to analyze. It can be shown that

$$P_{e,0} = P_{e,1} = 1 - \Phi\left(\sqrt{\frac{2A^2T}{N_0}}\right)$$

$$= 1 - \Phi\left(\sqrt{\frac{2\mathcal{E}_b}{N_0}}\right)$$

for the optimal receiver. For the same numerical example as before (i.e., $\mathcal{E}_b/N_0 = 5$), the error probabilities for the optimal receiver are

$$P_{e,0} = P_{e,1} = 1 - \Phi(\sqrt{10})$$
$$\approx 1 - \Phi(3.162)$$
$$\approx 0.00783 = 7.83 \times 10^{-4}. \qquad \blacksquare$$

PROBLEMS

3.1 A random process is defined by $X(t) = V p_T(t)$, where V is a random variable that is uniformly distributed on $[0, 1]$ and $p_T(t)$ is the rectangular pulse defined in Exercise 3–5. The random process is the input to a linear filter with impulse response $h(t) = p_{T_0}(t)$, and $Y(t)$ is the corresponding output.
  (a) Sketch some typical sample functions for the random process $X(t)$.
  (b) Sketch some typical sample functions for the output random process $Y(t)$ if $T_0 = T$.
  (c) Is the input random process wide-sense stationary?
  (d) Find the mean of $Y(t)$ if $T_0 > T$.
  (e) Find the autocorrelation function for the input random process $X(t)$.
  (f) Find $R_Y(t, t)$ if $T_0 = T$.

3.2 The zero-mean random process $Z(t)$ is Gaussian with autocorrelation function $R_Z(\tau) = \exp(-\tau^2)$. Find the probability that the sum $Z(0) + Z(1) + Z(2)$ exceeds the threshold γ. Express your answer in terms of γ, the exponential function, and the standard Gaussian distribution function Φ.

3.3 A discrete-time linear time-invariant filter has pulse response given by $h(n) = 1$ for $n = 0, 1$, and 2, and $h(n) = 0$ otherwise. The input is a discrete-time random process $X(n)$, $n \in \mathbb{Z}$, that has mean $\mu_X(n)$ and autocorrelation function $R_X(n, k) = \exp\{-(n - k)^2\}$. The corresponding output is the random process $Y(n)$, $n \in \mathbb{Z}$.
  (a) Find the mean $\mu_Y(n)$ of the output in terms of the mean of the input.
  (b) Assume that the input mean is identically zero, and find the crosscorrelation function $R_{X,Y}(i, j)$ in terms of the input autocorrelation function.
  (c) Assume that the input mean is identically zero, and find the output autocorrelation function $R_Y(i, j)$ in terms of the input autocorrelation function.
  (d) If the input mean is identically zero, are the processes $X(n)$ and $Y(n)$ jointly wide-sense stationary? Explain carefully why or why not.

3.4 The zero-mean random process $X(t)$ has autocorrelation function $R_X(\tau) = \beta \delta(\tau)$. $X(t)$ is the input to a time-invariant linear filter with impulse response given by

$$h(t) = \begin{cases} 1, & t_1 \leq t < t_2, \\ 0, & \text{otherwise}, \end{cases}$$

where $0 < t_1 < t_2$. Let $Y(t)$ be the corresponding output random process.
  (a) Are the input and output processes jointly wide-sense stationary?
  (b) Find the crosscorrelation function $R_{X,Y}(t, s)$ for the random processes $X(t)$ and $Y(t)$.
  (c) Find the autocorrelation function $R_Y(t, s)$ for the random process $Y(t)$.

3.5 Suppose $X(t)$ is a zero-mean wide-sense stationary continuous-time process with autocorrelation function $R_X(\tau) = \eta \exp\{-\gamma|\tau|\}$, $-\infty < \tau < \infty$. Let $X(t)$ be the input to the linear time-invariant filter with impulse response $h(t) = p_T(t)$. (Notice that this is the filter considered in Exercise 3–5.) Find the expected value of the instantaneous power in the output process at time t.

3.6 The random process $X(t)$ described in Problem 3.5 is the input to a simple RC filter with impulse response $h(t) = b \exp(-bt)$ for $t \geq 0$, where $b = 1/RC$. Assume that $1/RC \neq \gamma$. The corresponding output is $Y(t)$. Explain why the random processes $X(t)$ and $Y(t)$ are automatically jointly wide-sense stationary. Find the crosscorrelation function $R_{X,Y}(\tau)$.

3.7 Consider the same input random process and linear filter as in Problem 3.6. Find the output autocorrelation function $R_Y(\tau)$. The solution to Exercise 3–7 may be of help.

3.8 A continuous-time linear filter has impulse response $h(t) = p_\lambda(t)$. The input random process has the triangular autocorrelation function given as the second entry in Table 2–1 and illustrated in Figure 2–11(b). The output random process is $Y(t)$. Find the expected value of the instantaneous power in the output at time t_0. Consider three cases: $\lambda < T$, $\lambda = T$, and $\lambda > T$.

3.9 Suppose $X_1, X_2, X_3, \ldots$ is a sequence of independent, identically distributed random variables with distribution specified by

$$P(X_k = 0) = p, P(X_k = 1) = q, P(X_k = 2) = 1 - p - q,$$

and

$$P(X_k = n) = 0 \text{ for } n < 0 \text{ and for } n > 2.$$

Assume that p and q are positive numbers that satisfy the relation $p + q < 1$. From the sequence $X_1, X_2, X_3, \ldots$, we obtain another sequence $Z_1, Z_2, Z_3, \ldots$ by letting $Z_k = X_k + X_{k+1} + X_{k+2}$ for each positive integer k.
(a) Find $E\{Z_k\}$ and $E\{(Z_k)^2\}$ for each positive integer k.
(b) Find the autocorrelation $R_Z(k, k+1) = E\{Z_k Z_{k+1}\}$ for each positive integer k.
(c) Find the distribution for Z_k; that is, specify $P(Z_k = n)$ for each value of n and each positive integer k.
(d) Is the sequence $Z_1, Z_2, Z_3, \ldots$ a sequence of independent random variables? Answer this question by direct application of the definition of independent random variables.

3.10 The input to a linear filter with impulse response $p_T(t)$ is the sum of a deterministic signal $s(t) = \alpha p_T(t)$ and noise $X(t)$. Assume that the noise is wide-sense stationary with zero mean and autocorrelation function $R_X(\tau) = \beta \exp(-\tau^2/2)$. The output consists of the sum of $\hat{s}(t)$, the filtered version of the signal, and $\hat{X}(t)$, the filtered version of the noise. Suppose this output is sampled at time t_0, and the output signal-to-noise ratio is defined as $SNR(t_0) = \hat{s}(t_0)/\sigma$, where σ is the standard deviation of the noise $\hat{X}(t_0)$.
(a) Find the output signal-to-noise ratio as a function of t_0 (for $-\infty < t_0 < \infty$).
(b) Find the value of t_0 that maximizes the output signal-to-noise ratio, and find the resulting maximum value of $SNR(t_0)$.
(c) The input is sampled at time τ_0 $(0 < \tau_0 < T)$. Give the corresponding definition for the input signal-to-noise ratio.
(d) For the definition of the signal-to-noise ratio obtained in part (c), find the input signal-to-noise ratio.
(e) Compare the input signal-to-noise ratio and the maximum output signal-to-noise ratio from part (b), and show that for large values of T (i.e., $T \gg 1$), the filter improves the signal-to-noise ratio.

3.11 A discrete-time feedback control system has the property that its output voltage $X(k + 1)$ at time $k + 1$ is a linear combination of the output voltage $X(k)$ at time k and a random error $Y(k + 1)$ that is independent of past outputs. The equation that governs the system is

$$X(k + 1) = \alpha X(k) + Y(k + 1),$$

for each integer k, where $|\alpha| < 1$. Assume that the random process $Y(n)$, $n \in \mathbb{Z}$, is a sequence of independent, identically distributed random variables with mean zero and standard deviation β. Assume also that the random process $X(n)$, $n \in \mathbb{Z}$, has zero mean. The statement that the random error $Y(k + 1)$ is independent of past outputs means that $Y(k + 1)$ is independent of $\{X(n) : n \leq k\}$.

(a) Sketch a block diagram for this system.

(b) Is the output random process $X(n)$, $n \in \mathbb{Z}$, wide-sense stationary? Explain.

(c) Find the crosscorrelation function $R_{X,Y}(i, j)$ for all i and j.

(d) Find the autocorrelation $R_X(k, k + 1)$ for arbitrary k.

(e) Find the complete autocorrelation function for $X(n)$, $n \in \mathbb{Z}$. (It may be easier to first find $R_X(k, k + 2)$, $R_X(k, k + 3)$, etc., and observe the pattern.)

3.12 A time-invariant linear filter has impulse response given by $h(t) = t/T$ for $0 \leq t < T$ and $h(t) = 0$ otherwise, as shown in the following diagram:

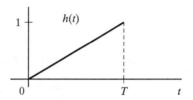

The input to this filter is a continuous-time random process $X(t)$, and the corresponding output is the random process $Y(t)$. If the input random process is white noise with autocorrelation function $R_X(\tau) = \beta\delta(\tau)$, find the autocorrelation function for the output process.

3.13 Consider the following system

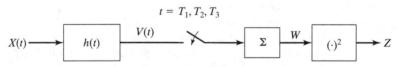

System model for Problem 3.13.

The input random process $X(t)$ is a Gaussian white-noise process that has spectral density $N_0/2$. The filter is a linear, time-invariant, causal filter with impulse response $h(t) = \exp(-t/T)$ for $0 \leq t < \infty$. The filter output is $V(t)$ when $X(t)$ is the input. The random variable W is given by

$$W = V(T_1) + V(T_2) + V(T_3),$$

and the sampling times are $T_1 = T$, $T_2 = 2T$, and $T_3 = 4T$. The system output is $Z = W^2$.

(a) Is $V(t)$ wide-sense stationary? Explain why or why not.

(b) Find the autocorrelation function for $V(t)$. Express your answer in terms of the parameters N_0 and T only. Give a mathematical expression for the autocorrelation, and also draw a sketch of its graph.

(c) Find $\mu = E\{Z\}$. Express your answer in terms of the parameters N_0 and T only.

(d) Find $P(Z \leq \gamma)$ for each choice of the real number γ. Express your answer in terms of the function $\Phi(\cdot)$ and the parameter μ of part **(c)**, not in terms of N_0 or T.

3.14 The continuous-time random process $X(t)$, $-\infty < t < \infty$, is wide-sense stationary and has autocorrelation function $R_X(\tau)$. The random process $Y(t)$, $-\infty < t < \infty$, is the output of a continuous-time, linear, time-invariant filter when $X(t)$ is the input.

(a) What facts about the filter and the random processes permit you to conclude that the input and output processes are *jointly* wide-sense stationary?

(b) Give the definition for the crosscorrelation function $R_{Y,X}(\tau)$ in terms of the expectation operator $E\{\ \}$ and the random processes $X(t)$, $-\infty < t < \infty$, and $Y(t)$, $-\infty < t < \infty$.

(c) Suppose the linear time-invariant filter has impulse response

$$h(t) = \begin{cases} 3, & t_1 \le t < t_2, \\ 0, & \text{otherwise}, \end{cases}$$

where $0 < t_1 < t_2$. Give an integral expression for $R_{Y,X}(\tau)$ with finite limits on the integral. Your answer should be in terms of τ, the parameters t_1 and t_2, and the autocorrelation function for the process $X(t)$, not in terms of the function h.

(d) Now suppose that the autocorrelation function for the process $X(t)$ is

$$R_X(\tau) = 7\exp(-5|\tau|), \quad -\infty < \tau < \infty.$$

Give a simple expression (e.g., evaluate all integrals) for $R_{Y,X}(\tau)$ for $\tau < 0$ only. Give your answer in terms of the parameters t_1 and t_2. A sketch of the functions involved may be of help.

(e) Is the correct answer to (d) also valid for some positive values of τ?

(f) If you said yes in part (e), give the range of positive values of τ for which the correct answer to (d) is valid. If you said no in part (e), explain in words or with a sketch why the correct answer to (d) is invalid for all $\tau > 0$.

Frequency-Domain Analysis of Random Processes in Linear Systems

4.0   THE USE OF FOURIER TRANSFORM TECHNIQUES

Just as they do in the analysis of deterministic signals in linear systems, Fourier transform techniques offer an alternative approach to the time-domain analysis of random processes in linear systems. This application of Fourier transform methods frequently leads to considerable computational savings and a better physical understanding of the problem at hand. The computational advantage stems from the relative ease of multiplying two Fourier transforms, compared with evaluating the convolution of two time-domain functions. The physical insight is derived from the ability to investigate the frequency distribution of the power in the random process, and such an investigation often provides important guidance in the design of filters to smooth the noise or improve the signal-to-noise ratio of a system.

There is one very important difference between the frequency-domain analysis of deterministic signals and the frequency-domain analysis of random processes: Just as in the time-domain analysis of random processes, we do not (and, in fact, cannot) apply Fourier transform methods to the individual sample functions of a random process. Instead, Fourier analysis techniques are applied primarily to the autocorrelation functions of the wide-sense stationary random processes involved. Although Fourier techniques can obviously be employed to evaluate the mean of the output process of a linear time-invariant system [i.e., to evaluate the convolution integral in equation (3.31)], this is not a very important application of Fourier techniques to the analysis of random processes. The most important applications deal with the Fourier transforms of the autocorrelation functions of wide-sense stationary random processes.

In this chapter, we introduce the spectral density function for wide-sense stationary random processes and present some of the properties of that function. We then develop the spectral analysis of random processes in linear systems and close the chapter (and the book) with discussions of the spectral density of a modulated signal and the representations of band-pass random processes and band-pass filters.

4.1   THE SPECTRAL DENSITY OF A WIDE-SENSE STATIONARY RANDOM PROCESS

The (power) *spectral density function* for a wide-sense stationary random process $X(t)$ with autocorrelation function R_X is defined by

$$S_X(\omega) = \int_{-\infty}^{\infty} R_X(\tau)e^{-j\omega\tau}\, d\tau, \tag{4.1}$$

whenever the integral exists. That is, *the spectral density function is the Fourier transform of the autocorrelation function.*

A few remarks about Fourier transforms are in order before we discuss the properties of spectral densities. In general, if g is a real-valued function of a real variable (i.e., $g : \mathbb{R} \to \mathbb{R}$), the Fourier transform of the function g is another function $G = \mathscr{F}\{g\}$ that is a complex-valued function of a real variable (i.e., $G : \mathbb{R} \to \mathbb{C}$ where $\mathbb{C}$ denotes the set of all complex numbers). The function $G = \mathscr{F}\{g\}$ is defined by

$$G(\omega) = \int_{-\infty}^{\infty} g(t)e^{-j\omega t}\, dt, \quad -\infty < \omega < \infty, \tag{4.2}$$

again, provided that the integral exists. The integral does exist whenever

$$\int_{-\infty}^{\infty} |g(t)|\, dt < \infty,$$

for instance, but this is not a necessary condition for defining the Fourier transform of the function g. We can consider Fourier transforms of functions such as $g(t) = \sin(2\pi t)$ by the use of generalized functions in the frequency domain (e.g., delta functions).

The preceding comments imply that the integral in (4.1) exists and the spectral density function S_X is well defined if the autocorrelation function satisfies

$$\int_{-\infty}^{\infty} |R_X(\tau)|\, d\tau < \infty, \tag{4.3}$$

even though this condition is not an absolute requirement. In applications to communications system design and analysis, it is often necessary to work with the spectral density function for a wide-sense stationary random process $X(t)$ whose autocorrelation function does *not* satisfy (4.3). There are many important autocorrelation functions, such as $R_X(\tau) = \cos(2\pi f_0\tau)$, that do not satisfy (4.3), but that can be handled by the use of generalized functions. In fact, the spectral density that corresponds to this autocorrelation function consists of a pair of delta functions in the frequency domain. (See Table 4–1 at the end of the section.)

The variable ω in (4.1) and (4.2) is usually referred to as the *radian frequency* and is measured in radians per second. The frequency $\omega/2\pi$ is the "usual" frequency—the number of cycles per second, measured in hertz (Hz). A standard abuse of notation is to write $S_X(f)$ for the spectral density as a function of frequency in Hz. A more precise notation is $S_X(2\pi f)$, but the notation $S_X(f)$ is more compact and more common in the literature. There is one important fact to remember in the use of the notation $S_X(f)$ and in the expression of spectral densities in terms of the frequency in Hz:

> *To convert an expression for the spectral density as a function of ω to an expression for the same spectral density as a function of f, we must replace ω by $2\pi f$, except in the argument of the function S_X itself.*

For example, if $S_X(\omega) = (1 + \omega^2)^{-1}$ for all ω, then $S_X(f) = (1 + 4\pi^2 f^2)^{-1}$ for all f. If $S_X(\omega) = N_0/2$ for all ω, then $S_X(f) = N_0/2$ for all f.

Formally, the spectral density as a function of frequency in Hz is defined as

$$S_X(f) = \int_{-\infty}^{\infty} R_X(\tau) e^{-j2\pi f\tau}\, d\tau.$$

A comparison of this expression with (4.1) shows that ω has been replaced by $2\pi f$ in the integrand, which is the basis for the conversion described in the previous paragraph.

Physically, $S_X(\omega)$ represents the density of power at frequency ω radians/sec, and $S_X(f)$ plays the same role for frequencies in Hz. The power in any frequency band is obtained by integrating the spectral density over the range of frequencies that make up the band. To develop the relationship between power and the integral of the spectral density function, the approach used here is to show that the inverse transform of the spectral density, when evaluated at the origin, gives the expected value of the instantaneous power in the random process.

Because $S_X = \mathcal{F}\{R_X\}$, the *inverse transform* of the spectral density is the auto-correlation function: $R_X = \mathcal{F}^{-1}\{S_X\}$. In integral form, this inverse transform can be written in either of the following two ways:

$$R_X(\tau) = \int_{-\infty}^{\infty} S_X(\omega) e^{+j\omega\tau}\, d\omega/2\pi$$

$$= \int_{-\infty}^{\infty} S_X(f) e^{+j2\pi f\tau}\, df.$$

In particular, the power in the random process $X(t)$ is

$$E\{[X(t)]^2\} = R_X(0) = \frac{1}{2\pi} \int_{-\infty}^{\infty} S_X(\omega)\, e^0\, d\omega$$

$$= \frac{1}{2\pi} \int_{-\infty}^{\infty} S_X(\omega)\, d\omega = \int_{-\infty}^{\infty} S_X(f)\, df.$$

This relationship shows that the power in the random process is the total area under the spectral density $S_X(f)$ of the process. Careful attention should be paid to the presence of the factor 2π in the integral of $S_X(\omega)$ and the absence of this factor in the integral of $S_X(f)$. The issue has a lot to do with the preference among many authors for working with $S_X(f)$ rather than $S_X(\omega)$. In general, the power in a frequency band from f_1 Hz to f_2 Hz is given by

$$\frac{1}{2\pi} \int_{2\pi f_1}^{2\pi f_2} S_X(\omega)\, d\omega = \int_{f_1}^{f_2} S_X(f)\, df.$$

This equation is a special case of some results developed in Section 4.2.

The properties of the spectral density function for a wide-sense stationary random process $X(t)$ that are most useful in frequency-domain analysis are as follows:

Property 1: $S_X(\omega) = S_X(-\omega)$, for all ω

Property 2: $S_X(\omega) = [S_X(\omega)]^*$, for all ω

Property 3: $S_X(\omega) \geq 0$, for all ω

Property 4:   If $\int_{-\infty}^{\infty} |R_X(\tau)|\, d\tau < \infty$, then $S_X(\omega)$ is a continuous function of ω.

In property 2, $[S_X(\omega)]^*$ denotes the *complex conjugate* of $S_X(\omega)$. The first three properties can be summarized by saying that a spectral density function is (1) even, (2) real, and (3) nonnegative. The first two of these properties are easily derived; the third and fourth are somewhat more difficult, and their derivations are postponed until Section 4.3.

To derive the formula $S_X(\omega) = S_X(-\omega)$, we make a change of variable $(\lambda = -\tau)$ and use the fact that $R_X(\lambda) = R_X(-\lambda)$. This procedure yields the following sequence of identities:

$$S_X(\omega) = \int_{-\infty}^{\infty} R_X(\tau)e^{-j\omega\tau}\, d\tau = \int_{-\infty}^{\infty} R_X(-\lambda)e^{+j\omega\lambda}\, d\lambda$$

$$= \int_{-\infty}^{\infty} R_X(\lambda)e^{-j(-\omega)\lambda}\, d\lambda = S_X(-\omega).$$

To prove that $S_X(\omega) = [S_X(\omega)]^*$, we can simply use the fact that $S_X(-\omega) = [S_X(\omega)]^*$ and then apply property 1. To see that $S_X(-\omega) = [S_X(\omega)]^*$ notice that

$$[S_X(\omega)]^* = \left\{ \int_{-\infty}^{\infty} R_X(\tau)\exp(-j\omega\tau)\, d\tau \right\}^* = \int_{-\infty}^{\infty} R_X(\tau)[\exp(-j\omega\tau)]^*\, d\tau$$

$$= \int_{-\infty}^{\infty} R_X(\tau)\exp(+j\omega\tau)\, d\tau = \int_{-\infty}^{\infty} R_X(\tau)e^{-j(-\omega)\tau}\, d\tau$$

$$= S_X(-\omega).$$

The two exercises that follow illustrate the conversion between autocorrelation functions and spectral density functions.

Exercise 4–1. Suppose the continuous-time random process $X(t)$ has autocorrelation function
$$R_X(\tau) = 1 + e^{-\alpha|\tau|}, \quad -\infty < \tau < \infty,$$
where $\alpha > 0$. Find the spectral density function for $X(t)$.

Solution. The Fourier transform of the constant 1 is $2\pi\delta(\omega)$. The Fourier transform of $e^{-\alpha|\tau|}$ is $2\alpha/(\alpha^2 + \omega^2)$. Hence,

$$S_X(\omega) = 2\pi\delta(\omega) + \frac{2\alpha}{\alpha^2 + \omega^2}. \qquad\blacksquare$$

Exercise 4–2. A zero-mean wide-sense stationary random process $X(t), -\infty < t < \infty$, has power spectral density

$$S_X(\omega) = \frac{1}{1 + \omega^2}, \quad -\infty < \omega < \infty.$$

Find the mean and variance of the random process defined by

$$Y(t) = \sum_{k=0}^{2} X(t + k).$$

Solution. The inverse transform of $S_X(\omega)$ is

$$R_X(\tau) = \tfrac{1}{2}e^{-|\tau|}.$$

Because $E\{Y(t)\} = 0$,

$$\text{Var}\{Y(t)\} = E\{[Y(t)]^2\}$$

$$= E\left\{\sum_{k=0}^{2}\sum_{i=0}^{2} X(t+k)X(t+i)\right\}$$

$$= \sum_{k=0}^{2}\sum_{i=0}^{2} R_X(k-i)$$

$$= 3R_X(0) + 2R_X(2) + 4R_X(1)$$

$$= \left(\tfrac{3}{2}\right) + 2e^{-1} + e^{-2}.$$

The solution to this exercise requires an understanding of *both* time- and frequency-domain descriptions, which is often the case for engineering problems that deal with wide-sense stationary random processes.   ∎

Recall that white noise has an autocorrelation function that is proportional to a delta function. Specifically, if $X(t)$ is a continuous-time white-noise process, its auto-correlation function is $R_X(\tau) = \frac{1}{2}N_0\delta(\tau)$. If we substitute this expression into (4.1), we find that *the spectral density function for the white noise process* is

$$S_X(\omega) = \frac{N_0}{2}, \quad -\infty < \omega < \infty.$$

Thus, white noise has a constant power spectral density over the entire frequency band. This spectral density function can also be written as

$$S_X(f) = \frac{N_0}{2}, \quad -\infty < f < \infty.$$

Table 4–1 gives a list of useful autocorrelation functions and their spectral density functions that arise frequently in engineering problems. The parameters α, T, and W that appear in the table are arbitrary positive constants. The list can be expanded easily by the use of simple identities such as the fact that, for an arbitrary positive constant c, the spectral density corresponding to $cR_X(\tau)$ is just $cS_X(\omega)$. Also, if $R_Z(\tau) = R_X(\tau) + R_Y(\tau)$, then $S_Z(\omega) = S_X(\omega) + S_Y(\omega)$.

Recall that multiplication in the time domain corresponds to convolution in the frequency domain. That is, if

$$g(t) = g_1(t)g_2(t),$$

and if $G_1 = \mathcal{F}\{g_1\}$, $G_2 = \mathcal{F}\{g_2\}$, and $G = \mathcal{F}\{g\}$, then

$$G(\omega) = \frac{1}{2\pi}\int_{-\infty}^{\infty} G_1(\alpha)G_2(\omega - \alpha)\, d\alpha, \quad -\infty < \omega < \infty, \tag{4.4a}$$

and

$$G(f) = \int_{-\infty}^{\infty} G_1(\beta)G_2(f - \beta)\, d\beta, \quad -\infty < f < \infty. \tag{4.4b}$$

It follows that if autocorrelation functions are related by

$$R_Z(\tau) = R_X(\tau)R_Y(\tau),$$

TABLE 4–1   Some Common Autocorrelation–Function/Spectral-Density Pairs[1]

	$R_X(\tau)$	$S_X(\omega)$
(1)	$\begin{cases}(T - \|\tau\|)/T, & \|\tau\| < T, \\ 0, & \text{otherwise}\end{cases}$	$T\{\sin(\omega T/2)/(\omega T/2)\}^2 = T\,\text{sinc}^2(fT)$
(2)	1	$2\pi\delta(\omega)$
(3)	$\delta(\tau)$	1
(4)	$\exp(-\alpha\|\tau\|)$	$2\alpha/(\alpha^2 + \omega^2)$
(5)	$\cos(\omega_0\tau)$	$\pi\delta(\omega - \omega_0) + \pi\delta(\omega + \omega_0)$
(6)	$\exp(-\alpha\|\tau\|)\cos(\omega_0\tau)$	$\{\alpha/[\alpha^2 + (\omega - \omega_0)^2]\} + \{\alpha/[\alpha^2 + (\omega + \omega_0)^2]\}$
(7)	$2W\,\text{sinc}(2W\tau) = \sin(2\pi W\tau)/\pi\tau$	$\begin{cases}1, & \|\omega\| \le 2\pi W \\ 0, & \text{otherwise}\end{cases}$

[1]Note the definition $\text{sinc}(u) = \sin(\pi u)/\pi u, -\infty < u < \infty$.

then the corresponding spectral densities are related by

$$S_Z(\omega) = \frac{1}{2\pi}(S_X * S_Y)(\omega) \tag{4.5a}$$

and

$$S_Z(f) = (S_X * S_Y)(f). \tag{4.5b}$$

For example, we can use (4.5a) to obtain entry (6) in Table 4–1 from entries (4) and (5). It is also possible to expand the list by using the fact that convolution in the time domain corresponds to multiplication in the frequency domain.

4.2   SPECTRAL ANALYSIS OF RANDOM PROCESSES IN LINEAR SYSTEMS

The main problem that we are concerned with in this section is the determination of the spectral density $S_Y(\omega)$ of the output $Y(t)$ of a time-invariant linear system for which the input is $X(t)$, a wide-sense stationary random process having spectral density $S_X(\omega)$. We let $h(t)$ be the impulse response of the system and $H(\omega)$ be the Fourier transform of $h(t)$ (i.e., $H = \mathcal{F}\{h\}$). The function $H(\omega)$ is called the *transfer function* of the system. Often, it is $H(\omega)$, rather than $h(t)$, that is specified in practice.

The first step in determining the spectral density is to notice that, if $\widetilde{h}$ is the "time-reverse impulse response" (defined in Chapter 3), the Fourier transform of $\widetilde{h}$ is

$$\widetilde{H}(\omega) = \int_{-\infty}^{\infty} h(-t)e^{-j\omega t}\,dt = \int_{-\infty}^{\infty} h(u)e^{+j\omega u}\,du$$

$$= \left[\int_{-\infty}^{\infty} h(u)e^{-j\omega u}\,du\right]^*$$

$$= [H(\omega)]^*,$$

where, as before, z^* denotes the complex conjugate of the complex number z.

Thus, $H(\omega)\tilde{H}(\omega) = H(\omega)[H(\omega)]^* = |H(\omega)|^2$. If we now use the fact that convolution in the time domain corresponds to multiplication in the frequency domain, we see that (3.51) corresponds to

$$S_Y(\omega) = H(\omega)\tilde{H}(\omega)S_X(\omega) = |H(\omega)|^2 S_X(\omega). \tag{4.6}$$

This is the key result for the spectral analysis of wide-sense stationary random processes in time-invariant linear systems. Several important facts about spectral densities can be deduced from (4.6), including the fact that the power (i.e., the expected value of the instantaneous power) in the output process is

$$E\{Y^2(t)\} = R_Y(0) = \int_{-\infty}^{\infty} |H(\omega)|^2 S_X(\omega) \, d\omega/2\pi. \tag{4.7}$$

This equation can also be written as

$$R_Y(0) = \int_{-\infty}^{\infty} |H(f)|^2 S_X(f) \, df. \tag{4.8}$$

Because of (4.7) and (4.8), $|H(\omega)|^2$ (or $|H(f)|^2$) is sometimes called the *power transfer function* for the linear filter. It is often easier to measure $|H(\omega)|^2$ than it is to measure $H(\omega)$. (The latter requires both phase and amplitude measurements.) If $H(\omega)$ or $|H(\omega)|^2$ is either specified or easily determined from the information that is given, then (4.6) is usually easier to work with than the corresponding time-domain expression, which is (3.51).

Exercise 4–3. Suppose the random process $X(t)$ of Exercise 4–1 is the voltage across a series *RLC* circuit. Let $Y(t)$ be the voltage across the capacitor. Find the spectral density function of the process $Y(t)$.

Solution. Since $Y(t)$ is the output of a linear system for which $X(t)$ is the input, it follows that

$$S_Y(\omega) = |H(\omega)|^2 S_X(\omega).$$

For the given system,

$$H(\omega) = [j\omega RC - \omega^2 LC + 1]^{-1}$$

and

$$|H(\omega)|^2 = [\omega^2 (RC)^2 + (1 - \omega^2 LC)^2]^{-1}.$$

Thus,

$$S_Y(\omega) = [\omega^2(RC)^2 + (1 - \omega^2 LC)^2]^{-1}\{2\pi\delta(\omega) + 2\alpha(\omega^2 + \alpha^2)^{-1}\}$$
$$= 2\pi\delta(\omega) + 2\alpha\{(\omega^2 + \alpha^2)\{(LC)^2\omega^4 + [(RC)^2 - 2LC]\omega^2 + 1\}\}^{-1}. \quad \blacksquare$$

If the input random process $X(t)$ is white noise with spectral density $N_0/2$, its spectral density function is defined by $S_X(\omega) = N_0/2, -\infty < \omega < \infty$. As a result, (4.6) simplifies to

$$S_Y(\omega) = \frac{N_0}{2}|H(\omega)|^2$$

Similarly, if the input process is white noise, (4.7) and (4.8) simplify to

$$R_Y(0) = \frac{N_0}{2} \int_{-\infty}^{\infty} |H(\omega)|^2 \, d\omega / 2\pi \tag{4.9}$$

and

$$R_Y(0) = \frac{N_0}{2} \int_{-\infty}^{\infty} |H(f)|^2 \, df, \tag{4.10}$$

respectively. From Chapter 3, we know that if $R_X(\tau) = \frac{1}{2} N_0 \delta(\tau)$, then

$$R_Y(0) = \frac{N_0}{2} \int_{-\infty}^{\infty} [h(t)]^2 \, dt. \tag{4.11}$$

[Equation (4.11) can be obtained by setting $\tau = 0$ in (3.81), for instance.] It follows from (4.9)–(4.11) that if the input to a time-invariant linear system is white noise, the power in the output process can be obtained by integrating either the square of the *magnitude* of the transfer function in the frequency domain or the square of the impulse response in the time domain. Some readers may notice that the equality between (4.10) and (4.11) can also be deduced from Parseval's relation.

The choice of which equation to use, (4.10) or (4.11), depends on the linear system under consideration. This is true in general in the choice between frequency-domain methods and time-domain methods. *Usually*, if $H(\omega)$ or $|H(\omega)|^2$ is either specified or easily determined, frequency-domain methods are easier than time-domain methods. On the other hand, if the time-domain functions are simple to describe mathematically, such as a rectangular or triangular pulse, time-domain methods are nearly always easier than frequency-domain methods. In particular, time-limited functions lend themselves to time-domain methods, while band-limited functions are usually easier to handle in the frequency domain.

4.3   DERIVATION OF PROPERTIES 3 AND 4

In this section, we derive the following two properties of the spectral density function for a wide-sense stationary random process $X(t)$:

Property 3: $S_X(\omega) \geq 0$, for all ω

Property 4:   If $\displaystyle\int_{-\infty}^{\infty} |R_X(\tau)| \, d\tau < \infty$, then $S_X(\omega)$ is a continuous function of ω.

There are two different approaches to deriving property 3. We include both derivations here, since both are instructive and illustrate two different analysis techniques. For the first approach, we start with the observation that for $T > 0$,

$$0 \leq E\left\{ \left| \int_0^T X(t) e^{-j\omega t} \, dt \right|^2 \right\} = E\left\{ \left[\int_0^T X(t) e^{-j\omega t} \, dt \right] \left[\int_0^T X(s) e^{-j\omega s} \, ds \right]^* \right\}$$

$$= E\left\{ \int_0^T X(t) e^{-j\omega t} \, dt \int_0^T X(s) e^{+j\omega s} \, ds \right\}$$

$$= E\left\{ \int_0^T \int_0^T X(t)X(s)e^{-j\omega(t-s)}\, dt\, ds \right\}$$

$$= \int_0^T \int_0^T R_X(t-s)e^{-j\omega(t-s)}\, dt\, ds.$$

Thus, this last double integral is nonnegative. Because $T > 0$, T^{-1} is also nonnegative, and we conclude that

$$T^{-1} \int_0^T \int_0^T R_X(t-s)e^{-j\omega(t-s)}\, dt\, ds \geq 0$$

for all positive values of T. Property 3 then follows from the fact that, for each ω and each $T > 0$,

$$\int_{-T}^T T^{-1}(T - |\tau|)R_X(\tau)e^{-j\omega\tau}\, d\tau = T^{-1} \int_0^T \int_0^T R_X(t-s)e^{-j\omega(t-s)}\, dt\, ds \qquad (4.12)$$

and

$$\lim_{T\to\infty} \int_{-T}^T T^{-1}(T - |\tau|)R_X(\tau)e^{-j\omega\tau}\, d\tau = S_X(\omega). \qquad (4.13)$$

Note that (4.12) and (4.13) establish that, for each ω, $S_X(\omega)$ is the limit of a sequence of nonnegative numbers, and the limit of a sequence of nonnegative numbers cannot be negative. Thus, property 3 is verified once we derive (4.12) and (4.13).

To derive (4.12), we use a change-of-variable transformation $F(t, s) = (t, t - s)$ and then substitute τ for $t - s$. The transformation F maps the original region of integration $\{(t, s) : 0 \leq t \leq T, 0 \leq s \leq T\}$ onto the region $G = \{(t, \tau) : t - T \leq \tau \leq t, 0 \leq t \leq T\}$. The original region is a square with vertices at the points $(0, 0)$, $(0, T)$, (T, T), and $(T, 0)$; the region G is a parallelogram with corner points $(0, 0)$, $(0, -T)$, $(T, 0)$, and (T, T). As illustrated in Figure 4–1, the region G can be written as

$$G = G_1 \cup G_2,$$

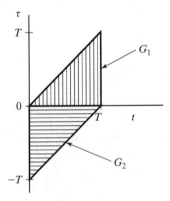

FIGURE 4–1   The region G.

where

$$G_1 = \{(t,\tau): \tau \leq t \leq T, 0 \leq \tau \leq T\} \text{ and } G_2 = \{(t,\tau): 0 \leq t \leq T+\tau, -T \leq \tau \leq 0\}.$$

From the decomposition of G into G_1 and G_2, it should be clear that the change-of-variable transformation permits the integral to be written as the sum of an integral over the region G_1 and an integral over the region G_2 as follows:

$$\int_0^T \int_0^T R_X(t-s)e^{-j\omega(t-s)} \, dt \, ds$$

$$= \int_0^T \left[\int_\tau^T R_X(\tau)e^{-j\omega\tau} \, dt \right] d\tau + \int_{-T}^0 \left[\int_0^{T+\tau} R_X(\tau)e^{-j\omega\tau} \, dt \right] d\tau$$

$$= \int_0^T (T-\tau)R_X(\tau)e^{-j\omega\tau} \, d\tau + \int_{-T}^0 (T+\tau)R_X(\tau)e^{-j\omega\tau} \, d\tau$$

$$= T \int_{-T}^T \frac{(T-|\tau|)}{T} R_X(\tau)e^{-j\omega\tau} \, d\tau.$$

This completes the derivation of (4.12).

To derive (4.13), note that if we define the function $\hat{R}_T$ by

$$\hat{R}_T(\tau) = \left[1 - \frac{|\tau|}{T}\right] R_X(\tau) \quad \text{for} \quad |\tau| \leq T$$

and $\hat{R}_T(\tau) = 0$ for $|\tau| > T$, then, for each fixed τ,

$$\lim_{T \to \infty} \hat{R}_T(\tau) = R_X(\tau).$$

Therefore,

$$\lim_{T \to \infty} \int_{-T}^T \left(1 - \frac{|\tau|}{T}\right) R_X(\tau)e^{-j\omega\tau} \, d\tau = \lim_{T \to \infty} \int_{-\infty}^\infty \hat{R}_T(\tau)e^{-j\omega\tau} \, d\tau$$

$$= \int_{-\infty}^\infty \lim_{T \to \infty} \hat{R}_T(\tau)e^{-j\omega\tau} \, d\tau$$

$$= \int_{-\infty}^\infty R_X(\tau)e^{-j\omega\tau} \, d\tau = S_X(\omega).$$

This derivation of (4.13) assumes that the order of the limit and the integration can be interchanged as in the foregoing sequence of steps. Because $|\hat{R}_T(\tau)| \leq |R_X(\tau)|$ for $T > 0$ and $-\infty < \tau < \infty$, this interchange is valid, for instance, if

$$\int_{-\infty}^\infty |R_X(\tau)| \, d\tau < \infty,$$

which is (4.3). The proof of the validity of the interchange follows from the dominated convergence theorem, which is stated and used shortly in the derivation of property 4. Before moving on to property 4, however, it is worth making two further observations about property 3. First, property 3 is actually valid for all processes of interest in engineering, even if (4.3) is not satisfied. However, the derivation of property 3 in the more general situation is more difficult than that just given. Second, we will show that property 3 also follows from (4.3) and property 4 (which is derived later).

The proof that (4.3) and property 4 imply property 3 is by contradiction. We suppose that (4.3) and property 4 hold. We then assert that property 3 is false and show that this leads to a contradiction. Because of our assertion that property 3 is false, there is at least one value of ω_0 for which $S_X(\omega_0) < 0$. By property 1, $S_X(-\omega_0) < 0$ also, so we can assume that $\omega_0 > 0$ if we wish. Since (4.3) is satisified, property 4 implies that $S_X(\omega)$ is *continuous*, so there must exist an ω_L and an ω_U such that $\omega_L < \omega_0 < \omega_U$ and $S_X(\omega) < 0$ for all ω in the interval (ω_L, ω_U). Let $Y(t)$ be the output of a band-pass filter $H(\omega)$ defined by letting $H(\omega) = 1$ for $|\omega|$ in the interval (ω_L, ω_U) and letting $H(\omega) = 0$ otherwise. Then

$$2\pi R_Y(0) = \int_{-\infty}^{\infty} S_Y(\omega)\, d\omega < 0,$$

which is a *contradiction* because $R_Y(0) \geq 0$ for any wide-sense stationary random process $Y(t)$. Hence, there cannot exist such an ω_0 as we have hypothesized; that is, it must be that $S_X(\omega) \geq 0$ for *all* ω. This proves that property 3 follows from (4.3) and property 4.

Notice that an *ideal* band-pass filter is not required in the preceding derivation. Any sufficiently close approximation suffices. All that is actually required of $H(\omega)$ is that

$$\int_{\omega_L}^{\omega_U} |S_X(\omega)||H(\omega)|^2\, d\omega > \int_{0}^{\omega_L} S_X(\omega)|H(\omega)|^2\, d\omega + \int_{\omega_U}^{\infty} S_X(\omega)|H(\omega)|^2\, d\omega.$$

The next step is to derive property 4. We make use of a result in mathematics known as the *dominated convergence theorem*, which gives one set of conditions under which it is valid to interchange the order of a limit and an integral. We have already used this theorem implicitly in establishing other results [e.g., (4.13)].

Theorem. Suppose that $f(x, y)$ is an integrable function of x for each y, and suppose also that

$$\lim_{y \to y_0} f(x, y) = f(x, y_0)$$

for each x. If there exists a function g such that

$$|f(x, y)| \leq g(x)$$

for each x and y, and if g satisfies

$$\int_{-\infty}^{\infty} g(x)\, dx < \infty,$$

then

$$\lim_{y \to y_0} \int_{-\infty}^{\infty} f(x, y)\, dx = \int_{-\infty}^{\infty} \lim_{y \to y_0} f(x, y)\, dx = \int_{-\infty}^{\infty} f(x, y_0)\, dx. \qquad \blacksquare$$

This theorem is applied to our problem by letting

$$f(\tau, \omega) = R_X(\tau) e^{-j\omega\tau}$$

and noticing that if $g(\tau) = |R_X(\tau)|$, then $|f(\tau, \omega)| \leq g(\tau)$ for each ω. Also, $f(\tau, \omega)$ is a continuous function of ω for each τ, so

$$\lim_{\omega \to \omega_0} f(\tau, \omega) = f(\tau, \omega_0) = R_X(\tau)e^{-j\omega_0\tau}.$$

Therefore,

$$\lim_{\omega \to \omega_0} S_X(\omega) = \lim_{\omega \to \omega_0} \int_{-\infty}^{\infty} R_X(\tau)e^{-j\omega\tau} \, d\tau = \lim_{\omega \to \omega_0} \int_{-\infty}^{\infty} f(\tau, \omega) \, d\tau.$$

An application of the dominated convergence theorem to the last expression establishes that

$$\lim_{\omega \to \omega_0} S_X(\omega) = \int_{-\infty}^{\infty} f(\tau, \omega_0) \, d\tau = \int_{-\infty}^{\infty} R_X(\tau)e^{-j\omega_0\tau} \, d\tau = S_X(\omega_0).$$

Thus, we have shown that, for any ω_0,

$$\lim_{\omega \to \omega_0} S_X(\omega) = S_X(\omega_0),$$

which is just the definition of continuity for the function S_X at the point ω_0.

Now for a word of warning: Although the requirement

$$\int_{-\infty}^{\infty} |R_X(\tau)| \, d\tau < \infty,$$

is not essential for several of our results (e.g., property 3), it is critical to property 4. In fact, it is easy to find examples of random processes for which

$$\int_{-\infty}^{\infty} |R_X(\tau)| \, d\tau = \infty,$$

and the corresponding spectral densities have discontinuities. For instance, if

$$R_X(\tau) = 2W \left[\frac{\sin 2\pi W\tau}{2\pi W\tau} \right],$$

then $S_X(\omega) = 1$ for $-2\pi W \leq \omega \leq 2\pi W$ and $S_X(\omega) = 0$ otherwise. This spectral density function has discontinuities at $-2\pi W$ and $+2\pi W$. The example $R_X(\tau) = \cos \tau$ is even worse: *Its spectral density consists of delta functions.*

4.4   SPECTRUM OF AMPLITUDE-MODULATED SIGNALS

Consider a signal of the form

$$Y(t) = \sqrt{2}A(t)\cos(\omega_c t + \Theta), \tag{4.14}$$

where $A(t)$ is a random process representing the amplitude modulation and Θ is a random variable representing the phase of the sinusoidal carrier upon which $A(t)$ is modulated. The carrier signal is $\sqrt{2}\cos(\omega_c t + \Theta)$. For analog communications, $A(t)$ may represent a speech signal. Such a signal is very noiselike, so for many applications it is best modeled as a random process. In a digital communication system, $A(t)$ is typically a continuous-time waveform that represents a sequence of data pulses, and there is randomness in the data sequence represented by the sequence of amplitudes of these pulses. In addition, there may be a random time shift in the sequence of pulses, which represents the delay in the communication channel or the random starting time of the transmission.

In nearly all systems, the phase angle Θ and the information bearing-signal $A(t)$ originate from different physical mechanisms, and it is appropriate to model them as statistically independent. In all that follows, we assume that Θ and $A(t)$ are independent, which implies that

$$E\{Y(t)\} = \sqrt{2}E\{A(t)\}E\{\cos(\omega_c t + \Theta)\}$$

and

$$E\{Y(t + \tau)Y(t)\} = 2E\{A(t + \tau)A(t)\}E\{\cos[\omega_c(t + \tau) + \Theta]\cos(\omega_c t + \Theta)\}.$$

We also assume that $A(t)$ is a wide-sense stationary process with autocorrelation function

$$R_A(\tau) = E\{A(t + \tau)A(t)\}.$$

By the trigonometric identity for the product of the cosines, we see that

$$E\{Y(t + \tau)Y(t)\} = R_A(\tau)E\{\cos(\omega_c\tau) + \cos[\omega_c(2t + \tau) + 2\Theta]\}. \quad (4.15)$$

As is always true for phase angles, the random variable Θ appears inside a trigonometric function, so we can subtract multiples of 2π from Θ without changing its effective value. Thus, there is no harm in limiting the range of Θ to the interval $[0, 2\pi]$. This is accomplished formally by considering the phase angle modulo 2π: If φ' takes values in the interval $-\infty < \varphi' < \infty$ and φ takes values in the interval $0 \le \varphi < 2\pi$, then φ' is equal to φ modulo 2π if $\varphi = \varphi' + 2\pi n$ for some choice of the integer n.

For most applications, the distribution for the random variable Θ should be modeled as the uniform distribution on the interval $[0, 2\pi]$. If Θ is so distributed, then

$$E\{\cos(\omega_c t + \Theta)\} = 0 \quad (4.16a)$$

and

$$E\{\cos[\omega_c(2t + \tau) + 2\Theta]\} = 0. \quad (4.16b)$$

Equation (4.16a) implies that $E\{Y(t)\} = 0$ for all t. Because the term $\cos(\omega_c\tau)$ is deterministic, it follows from (4.15) and (4.16b) that

$$E\{Y(t + \tau)Y(t)\} = R_A(\tau)\cos(\omega_c\tau).$$

We conclude that if $A(t)$ is wide-sense stationary, then $Y(t)$ is a zero-mean, wide-sense stationary random process with autocorrelation function

$$R_Y(\tau) = R_A(\tau)\cos(\omega_c\tau). \quad (4.17)$$

The *modulation theorem of Fourier transforms* states that if $v(t)$ has Fourier transform $V(f)$, then the Fourier transform of the signal

$$w(t) = v(t)\cos(\omega_c t) \quad (4.18a)$$

is

$$W(f) = \frac{V(f - f_c) + V(f + f_c)}{2},\qquad(4.18b)$$

where $f_c = \omega_c/2\pi$. The theorem is easily proved by using the fact that multiplication in the time domain corresponds to convolution in the frequency domain [e.g., see (4.4)]. The proof is completed by noting that the Fourier transform of $\cos(2\pi f_c t)$ is

$$\frac{\delta(f - f_c) + \delta(f + f_c)}{2}$$

and that the convolution of $\delta(f \pm f_c)$ with $V(f)$ is $V(f \pm f_c)$. An alternative proof is outlined in Problem 4.6.

   If the modulation theorem is applied to (4.17), we find that the spectral density for $Y(t)$ is given by

$$S_Y(f) = \frac{S_A(f - f_c) + S_A(f + f_c)}{2},\qquad(4.19)$$

where $S_A(f)$ denotes the spectral density for $A(t)$.

Example 4–1   A Random Data Signal with Rectangular Pulses

   We consider one model for a random sequence of rectangular pulses in which each pulse has unit amplitude and duration T. Thus, the basic waveform is the rectangular pulse defined by

$$p_T(t) = \begin{cases} 1, & 0 \le t < T, \\ 0, & \text{otherwise.} \end{cases}$$

Define a random process by

$$D(t) = \sum_{n=-\infty}^{\infty} A_n p_T(t - nT),$$

where A_n is a random variable that represents the amplitude of the nth pulse. Note that the pulses do not overlap: The nth pulse is nonzero only in the interval $nT \le t < (n + 1)T$. Even if the random variables A_n, $-\infty < n < \infty$, are independent and identically distributed, the random process $D(t)$ is still not wide-sense stationary. To see this, simply compare $E\{D(T/4)D(3T/4)\}$ with $E\{D(3T/4)D(5T/4)\}$. The former is $E\{A_0^2\}$ while the latter is $E\{A_0 A_1\} = E\{A_0\}E\{A_1\}$. Suppose, for example, that for each n, the mean of A_n is zero and the variance of A_n is unity. Then

$$E\left\{D\left(\frac{T}{4}\right)D\left(\frac{3T}{4}\right)\right\} = 1,$$

but

$$E\left\{D\left(\frac{3T}{4}\right)D\left(\frac{5T}{4}\right)\right\} = 0.$$

Similarly, $E\{D(T/4)D(T/2)\} = 1$, while $E\{D(-T/8)D(T/8)\} = 0$.   ■

   Example 4–1 illustrates the need for a random time offset between the data signal and the carrier if it is desired to model the amplitude modulation signal as a wide-sense stationary random process. Let the data pulse waveform be represented by the function ζ. For instance, it might be that $\zeta(t) = p_T(t)$, the rectangular pulse of Example 4–1. Suppose that

$$D(t) = \sum_{n=-\infty}^{\infty} A_n \zeta(t - nT), \qquad (4.20)$$

as in that example, and the signal

$$A(t) = D(t - U), \quad -\infty < t < \infty,$$

where U is a random variable that is uniformly distributed on the interval $[0, T]$. The variable U is a random time offset, which may be used to account for a random starting time or propagation delay between the transmitter and receiver. As long as the random variables $A_n, -\infty < n < \infty$, are independent and identically distributed, the range of U can be restricted to the interval $[0, T]$ for essentially the same reason that phase angles can be restricted to the interval $[0, 2\pi]$. Moreover, only minor changes are required in the development that follows if U is uniformly distributed on any interval of the form $[n_1 T, n_2 T]$ for arbitrary integers n_1 and n_2 (with $n_1 < n_2$, of course).

Suppose that the data sequence $A_n, -\infty < n < \infty$, is a sequence of independent random variables with $E\{A_n\} = 0$ and $E\{A_n^2\} = \alpha^2$ for each n. First, notice that $E\{A_n\} = 0$ for each n implies that

$$E\{A(t)\} = E\{D(t)\} = 0$$

for each t. In particular, the random process $A(t)$ has a constant mean. Next, notice that because the random variables $A_n, -\infty < n < \infty$, are independent, they are uncorrelated. Also, because the mean of each A_n is zero,

$$E\{A_n A_k\} = 0$$

for $n \neq k$.

The next step is to evaluate $E\{A(t + \tau)A(t)\}$ and show that it does not depend on t. This, together with the fact that $A(t)$ has a constant mean, implies that $A(t)$ is a wide-sense stationary random process, so that we can consider its spectral density.

For any t and τ,

$$E\{A(t + \tau)A(t)\} = E\left\{ \sum_{n=-\infty}^{\infty} A_n \zeta(t + \tau - nT - U) \sum_{k=-\infty}^{\infty} A_k \zeta(t - kT - U) \right\}$$

$$= \sum_{n=-\infty}^{\infty} \sum_{k=-\infty}^{\infty} E\{A_n A_k\} E\{\zeta(t + \tau - nT - U)\zeta(t - kT - U)\}$$

$$= \sum_{n=-\infty}^{\infty} \alpha^2 E\{\zeta(t + \tau - nT - U)\zeta(t - nT - U)\}. \qquad (4.21)$$

The last step follows from the fact that $E\{A_n A_k\} = 0$ for $n \neq k$. Because the random variable U is uniformly distributed on the interval $[0, T]$,

$$E\{\zeta(t + \tau - nT - U)\zeta(t - nT - U)\} = T^{-1} \int_0^T \zeta(t + \tau - nT - u)\zeta(t - nT - u)\, du$$

$$= T^{-1} \int_{nT}^{(n+1)T} \zeta(t + \tau - v)\zeta(t - v)\, dv.$$

The last step follows from the change of variable $v = u + nT$. Substituting into (4.21) gives

$$E\{A(t + \tau)A(t)\} = \sum_{n=-\infty}^{\infty} \alpha^2 T^{-1} \int_{nT}^{(n+1)T} \zeta(t + \tau - v)\zeta(t - v)\, dv$$

$$= \alpha^2 T^{-1} \sum_{n=-\infty}^{\infty} \int_{nT}^{(n+1)T} \zeta(t + \tau - v)\zeta(t - v)\, dv$$

$$= \alpha^2 T^{-1} \int_{-\infty}^{\infty} \zeta(t + \tau - v)\zeta(t - v)\, dv$$

$$= \alpha^2 T^{-1} \int_{-\infty}^{\infty} \zeta(u)\zeta(u - \tau)\, du. \tag{4.22}$$

The last step is just the change of variable $u = t + \tau - v$. Note in particular that

$$E\{A^2(t)\} = \alpha^2 T^{-1} \int_{-\infty}^{\infty} \zeta^2(u)\, du.$$

For $\zeta(t) = p_T(t)$, this gives $E\{A^2(t)\} = \alpha^2$.

It follows from (4.22) that $E\{A(t + \tau)A(t)\}$ does not depend on t, so that the random process $A(t)$ is a zero-mean, wide-sense stationary random process with autocorrelation function

$$R_A(\tau) = \alpha^2 T^{-1} \int_{-\infty}^{\infty} \zeta(u)\zeta(u - \tau)\, du. \tag{4.23}$$

This expression can also be written as

$$R_A(\tau) = \alpha^2 T^{-1}(\zeta * \tilde{\zeta})(\tau), \tag{4.24}$$

where $\tilde{\zeta}(t) = \zeta(-t)$, $-\infty < t < \infty$. That is, the function $\tilde{\zeta}$ is the time reverse of the pulse function ζ, and R_A is proportional to the convolution of the pulse waveform and its time reverse. Note the similarity between (4.24) and the results for time-invariant linear filtering of random processes [e.g., (3.53)].

Let Z denote the Fourier transform of ζ. We know that the Fourier transform of $\tilde{\zeta}$ is Z^*, the complex conjugate of Z. (This property is derived in Section 4.2.) In a manner analogous to the relationship between (3.51) and (4.6), it follows from (4.24) that the spectral density function for the random process $A(t)$ is given by

$$S_A(f) = \alpha^2 T^{-1}|Z(f)|^2. \tag{4.25}$$

This result can be employed in (4.19) to show that the spectral density function for the amplitude-modulated signal

$$Y(t) = \sqrt{2} A(t) \cos(\omega_c t + \Theta)$$

is

$$S_Y(f) = \frac{\alpha^2 \{|Z(f - f_c)|^2 + |Z(f + f_c)|^2\}}{2T}. \tag{4.26}$$

Example 4–2   A Random Data Signal with Rectangular Pulses (Revisited)

Consider a signal of the form

$$Y(t) = \sqrt{2}\, D(t - U) \cos(\omega_c t + \Theta), \tag{4.27}$$

where

$$D(t) = \sum_{n=-\infty}^{\infty} A_n p_T(t - nT), \tag{4.28}$$

as in Example 4–1. Notice that $D(t)$ is just a sequence of rectangular pulses of duration T. The amplitudes of these pulses are determined by the sequence A_n, $-\infty < n < \infty$. If we ignore the random time delay for the moment (i.e., temporarily set $U = 0$), then, for each value of t in the range $nT \le t < (n + 1)T$,

$$Y(t) = \sqrt{2}\, A_n \cos(\omega_c t + \Theta).$$

Information can be conveyed in the sequence of amplitudes of this signal, which is a common method in digital communications. We are interested in determining the spectral density of $Y(t)$ if U is modeled as a random variable that is uniformly distributed on $[0, T]$, Θ is uniformly distributed on $[0, 2\pi]$, and the sequence of amplitudes is a sequence of independent random variables satisfying $E\{A_n\} = 0$ and $E\{A_n^2\} = \alpha^2$ for each n. The random variables U, Θ, and A_n, $-\infty < n < \infty$, are assumed to be mutually independent. The determination of the spectral density for $Y(t)$ is accomplished by an application of (4.26) with $\zeta(t) = p_T(t)$.

There are two ways to proceed. We can use the fact that $P_T(f)$, the Fourier transform of the unit-amplitude rectangular pulse of duration T, satisfies

$$|P_T(f)| = |(\pi f)^{-1} \sin(\pi f T)|, \tag{4.29}$$

so that for $\zeta(t) = p_T(t)$,

$$|Z(f)|^2 = (\pi f)^{-2} \sin^2(\pi f T).$$

The second approach is to use the fact that if $\zeta(t) = p_T(t)$, the function $\zeta * \tilde{\zeta}$ is a triangular function centered at the origin with base $2T$ and height T. In fact, $(\zeta * \tilde{\zeta})(\tau) = f(\tau)$, the function derived in Exercise 3–5 and illustrated in Figure 3–4. We can then use the fact that the Fourier transform of such a triangular pulse is $(\pi f)^{-2} \sin^2(\pi f T)$, the same result as that obtained in the first approach. Using the fact that $\mathrm{sinc}(x) = (\pi x)^{-1} \sin(\pi x)$, we see that $|Z(f)|^2$ can be written more compactly as

$$|Z(f)|^2 = T^2 \mathrm{sinc}^2(fT). \tag{4.30}$$

It follows from (4.25) that the spectral density for $A(t)$ is

$$S_A(f) = \alpha^2 T \, \mathrm{sinc}^2(fT),$$

and it follows from (4.26) that the spectral density for $Y(t)$ is

$$S_Y(f) = \frac{\alpha^2 T \{ \mathrm{sinc}^2[(f - f_c)T] + \mathrm{sinc}^2[(f + f_c)T] \}}{2} \tag{4.31}$$

■

The spectrum of (4.31) can best be illustrated by taking advantage of the fact that its shape is derived from the spectrum

$$|Z(f)|^2 = T^2 \, \mathrm{sinc}^2(fT).$$

It is convenient to normalize $|Z(f)|^2$ by dividing it by T^2 and then plot the normalized spectrum as a function of fT. Because the units of f are inverse seconds (Hz) and the units of T are seconds, fT is a dimensionless parameter. A graph of $T^{-2}|Z(f)|^2$ as a function of the normalized frequency is shown in Figure 4–2 for $0 \le fT \le 3$. This spectrum is an even function of fT, so there is no need to show the graph for negative frequencies.

It is interesting to compare the results we have obtained by considering the communication signal as a random process with the corresponding results that are obtained if the signal is modeled as a deterministic signal. The key step in the latter approach is to use the energy spectrum for the deterministic signal. If the Fourier transform of the deterministic signal $f(t)$ is denoted by $F(\omega)$, then the *energy spectrum* of the signal is defined to be $|F(\omega)|^2$. Thus, the energy spectrum for the pulse waveform $\zeta(t)$ is $|Z(f)|^2$. Notice that the shape of the spectral density given in (4.26) depends only on the energy spectrum of the pulse waveform $\zeta(t)$ and the carrier frequency ω_c. This relationship is pursued further in Problem 4.8.

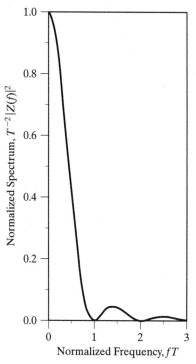

FIGURE 4–2   Spectral density for a random data signal with a rectangular pulse waveform.

4.5   BAND-PASS FREQUENCY FUNCTIONS

Recall that if V is the Fourier transform of v, then

$$V(\omega) = \int_{-\infty}^{\infty} v(t)e^{-j\omega t}\, dt \tag{4.32a}$$

and

$$v(t) = \int_{-\infty}^{\infty} V(\omega)e^{+j\omega t}\frac{d\omega}{2\pi}, \tag{4.32b}$$

where the variable ω represents frequency in radians per second. In terms of the frequency variable f in hertz, these equations become

$$V(f) = \int_{-\infty}^{\infty} v(t)e^{-j2\pi ft}\, dt \tag{4.33a}$$

and

$$v(t) = \int_{-\infty}^{\infty} V(f)e^{+j2\pi ft}\, df. \tag{4.33b}$$

In this section, much of the development is in terms of inverse Fourier transforms, for which it is convenient to use transform expressions (4.33) in order to avoid the need to carry along the factor 2π. Thus, all Fourier transforms in this section are given in terms of the variable f, the frequency in hertz.

The first result that we need follows easily from (4.33a): If $v(t)$ is real for each t, and V is the Fourier transform of v, then

$$V^*(f) = \left\{ \int_{-\infty}^{\infty} v(t)e^{-j2\pi ft}\, dt \right\}^*$$

$$= \left\{ \int_{-\infty}^{\infty} v(t)e^{+j2\pi ft}\, dt \right\} = V(-f). \tag{4.34}$$

Additional steps are given for this derivation in Section 4.1, where it is shown that $[S_X(\omega)]^* = S_X(-\omega)$ for each ω. Recall that S_X is the Fourier transform of the function R_X, and $R_X(\tau)$ is real for each τ. We refer to the property wherein $V^*(f) = V(-f)$ as *conjugate symmetry*, so (4.34) is summarized by saying that the Fourier transform of a real function has conjugate symmetry. If a time function is real, its Fourier transform has conjugate symmetry.

4.5.1   Alternative Definitions of the Bandwidth for Frequency Functions

The time-domain functions of interest in this section not only are real-valued functions, but also have Fourier transforms that are band limited in a special way. Suppose that W is a function that represents a frequency-domain mathematical description of some entity such as a signal, filter, or noise process. For example, W might be the transfer function for a time-invariant linear system, or it might be the spectral density for a wide-sense stationary random process. Because the function W characterizes the entity in the frequency domain, we refer to W as a *frequency function*. Its inverse transform is a real function that characterizes the entity in the time domain. Examples include the impulse response for a time-invariant linear filter and the autocorrelation function for a random process.

Although all of the illustrations in this section are for real frequency functions, all of the analysis is for complex-valued functions. Of course, this analysis applies to real frequency functions as a special case. Spectral densities are real functions, but transfer functions for filters of interest in applications are usually complex valued. A complex-valued frequency function can always be written as

$$W(f) = A(f) \exp\{j\Psi(f)\},$$

where A and Ψ are real frequency functions that represent the amplitude and phase, respectively, of the frequency function W. Notice that W has conjugate symmetry if and only if $A(f) = A(-f)$ and $\Psi(f) = -\Psi(-f)$ for all f.

A complete illustration of a complex-valued frequency function requires the display of both its amplitude and phase functions. Often, the amplitude function is of greater interest than the phase function and is the only one of the two that is displayed. One of the features that we wish to examine is the bandwidth of the frequency function, a parameter that depends only on the amplitude function. Our illustrations of real frequency functions can be viewed as illustrations of the amplitude functions for complex-valued frequency functions.

If there are two frequencies f' and f'' such that $0 < f' < f''$ and for which $W(f) \approx 0$ for all f outside the intervals $[-f'', -f']$ and $[f', f'']$, the frequency function W is referred to as a *band-pass frequency function*. If $W(f) = 0$ for all frequencies f outside the intervals $[-f'', -f']$ and $[f', f'']$, W is referred to as an *ideal band-pass frequency function*. The frequencies f' and f'' are not unique, because it is always possible to increase f'' and decrease f'. For the ideal band-pass frequency functions that are considered in this book, it is possible to let f_1 be the largest value of f' and f_2 the smallest value of f'' for which $W(f) = 0$ for all frequencies f outside the intervals $[-f'', -f']$ and $[f', f'']$. The resulting interval $[f_1, f_2]$ is referred to as the *frequency support* for W. An example of a band-pass frequency function is illustrated in Figure 4–3.

If the ideal band-pass frequency function W is a continuous function, the frequencies f_1 and f_2 can be defined mathematically by

$$f_1 = \max\{u : W(f) = 0 \text{ for } 0 \le f \le u\} \tag{4.35a}$$

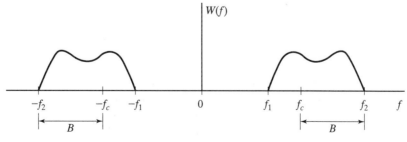

FIGURE 4–3   An ideal band-pass frequency function.

and

$$f_2 = \min\{v : W(f) = 0 \text{ for } f \geq v\}. \tag{4.35b}$$

Continuity guarantees that the maximum and minimum in (4.35) exist. The frequencies f_1 and f_2 represent the lower and upper *cutoff frequencies*, respectively, for the ideal band-pass frequency function W. The difference $f_2 - f_1$ is referred to as the *absolute bandwidth* or *cutoff bandwidth* of the band-pass frequency function. Thus, the absolute bandwidth is just the width of the frequency support for W. Notice that because W has conjugate symmetry, its cutoff frequencies and its bandwidth can be defined by considering positive frequencies only.

Other measures of bandwidth are useful in characterizing the spectral occupancy of signals and random processes. We mention here only two of the many alternative definitions. The first is the half-power bandwidth for a filter with transfer function W. Let W_m be the maximum value of $|W(f)|$, $-\infty < f < \infty$; this represents the maximum gain of the filter across the frequency band. The transfer function W has *half-power bandwidth* $f_4 - f_3$ if $0 < f_3 < f_4$ and if each of the following is true:

$$|W(f_3)| = |W(f_4)| = \frac{W_m}{\sqrt{2}},$$

$$|W(f)| > \frac{W_m}{\sqrt{2}} \quad \text{for } f_3 < f < f_4,$$

and

$$|W(f)| < \frac{W_m}{\sqrt{2}} \quad \text{for } 0 < f < f_3 \text{ and } f_4 < f < \infty.$$

Notice that a transfer function may not have a well-defined half-power bandwidth. (See Problem 4.11.) If the half-power bandwidth exists, then the frequencies f_3 and f_4 are such that the power transfer function $|W(f)|^2$ satifies

$$|W(f_3)|^2 = |W(f_4)|^2 = \frac{(W_m)^2}{2}.$$

If W represents a power spectral density, the conditions for f_3 and f_4 become

$$W(f_3) = W(f_4) = \frac{W_m}{2}.$$

Because $10 \log_{10}\left(\frac{1}{2}\right) \approx -3$ dB, the half-power bandwidth is often referred to as the 3-dB bandwidth.

Another bandwidth of interest is the *null-to-null bandwidth*. Under the following conditions, $f_6 - f_5$ is the null-to-null bandwidth of the frequency function W: f_5 and f_6 are such that $0 < f_5 < f_6$, the maximum value of $W(f)$ occurs between frequencies f_5 and f_6, $W(f) \neq 0$ for $f_5 < f < f_6$, and $W(f_5) = W(f_6) = 0$. For the frequency function illustrated in Figure 4–3, the null-to-null bandwidth is equal to $f_2 - f_1$, the same as the absolute bandwidth. As another example, it is easy to show that the null-to-null bandwidth of the spectral density

$$S_Y(f) = \frac{\alpha^2 T \{\mathrm{sinc}^2[(f - f_c)T] + \mathrm{sinc}^2[(f + f_c)T]\}}{2}$$

is $2/T$. This spectral density is derived in Example 4–2.

4.5.2 Time-Domain Descriptions of Ideal Band-Pass Frequency Functions

In this section, we seek the time-domain characterization of an ideal band-pass frequency function W that has conjugate symmetry (i.e., W is the Fourier transform of a real function). It is shown that if f_c is a suitable frequency, then the inverse Fourier transform of W can be written in the form

$$w(t) = v_1(t) \cos(2\pi f_c t) + v_2(t) \sin(2\pi f_c t),$$

where v_1 and v_2 are baseband functions.

The choice of the frequency f_c depends on the application. If $w(t)$ represents a carrier-modulated communication signal, f_c is almost always the carrier frequency of the signal. If $w(t)$ is the impulse response of a band-pass filter, f_c might be selected to be the center frequency for this filter, or it might be the carrier frequency of a signal that is the input to the filter. Under normal circumstances, f_c is located somewhere near the middle of the frequency support of the frequency function W, but this is not a requirement. All that is required is that f_c be within the frequency support of W; that is, if W has lower and upper cutoff frequencies f_1 and f_2, respectively, then $f_1 < f_c < f_2$. In this section, it is shown that for such a choice of f_c, if the absolute bandwidth of W is not greater than $2B$, then v_1 and v_2 have Fourier transforms that are identically zero for $|f| \geq B$.

If W is an ideal band-pass frequency function with lower and upper cutoff frequencies f_1 and f_2, respectively, and if f_c is between f_1 and f_2, then B can be defined by

$$B = \max\{f_2 - f_c, f_c - f_1\}.$$

That is, B is the frequency separation between f_c and the cutoff frequency that is farthest from f_c. It follows that $W(f) = 0$ for all f outside the intervals $[-f_c - B, -f_c + B]$ and $[f_c - B, f_c + B]$, so the absolute bandwidth is not greater than $2B$. The relationship that exists among f_1, f_2, f_c, and B is illustrated in Figure 4–3.

For the mathematical development that follows, we concentrate on ideal band-pass frequency functions. In practice, however, it is usually sufficient if $W(f)$ is *approximately* zero outside the intervals $[-f_2, -f_1]$ and $[f_1, f_2]$ in order that the time and frequency functions have the key properties derived in this section. Under this weaker condition, v_1 and v_2 have Fourier transforms that are approximately zero for $|f| \geq B$.

Let W_+ be the *positive part* of the band-pass frequency function W; that is,

$$W_+(f) = \begin{cases} W(f), & f > 0, \\ 0, & f \leq 0. \end{cases} \tag{4.36}$$

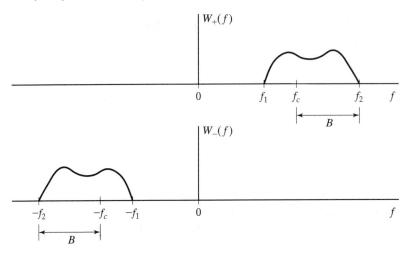

FIGURE 4–4   The positive and negative parts of W.

The positive part of W is illustrated in Figure 4–4. Notice that the frequency support for the positive part of W is a subset of the interval $(0, \infty)$.

Similarly, let W_- be the *negative part* of W; that is, $W_-(f) = W(f)$ for $f < 0$ and $W_-(f) = 0$ for $f \geq 0$. The negative part of W is also illustrated in Figure 4–4. The frequency support for the negative part of W is a subset of the interval $(-\infty, 0)$. The band-pass frequency function W can be written as

$$W(f) = W_+(f) + W_-(f).$$

The frequency support for W_+ and the frequency support for W_- are disjoint for a band-pass frequency function. That is, for each frequency f, either $W_+(f) = 0$ or $W_-(f) = 0$. This can be expressed concisely in terms of the product of the two functions:

$$W_+(f)W_-(f) = 0, \quad -\infty < f < \infty.$$

Because $W(f)$ is the Fourier transform of a real function, it has conjugate symmetry: $W^*(f) = W(-f)$. But this implies that $[W_+(f)]^* = W_-(-f)$, which guarantees that the inverse Fourier transform of W can be written in terms of the function W_+ only. The inverse transform of W is given by

$$w(t) = \int_{-\infty}^{\infty} W(f)e^{+j2\pi ft}\, df$$

$$= \int_{-\infty}^{\infty} W_+(f)e^{+j2\pi ft}\, df + \int_{-\infty}^{\infty} W_-(f)e^{+j2\pi ft}\, df. \qquad (4.37)$$

But

$$\int_{-\infty}^{\infty} W_-(f)e^{+j2\pi ft}\, df = \int_{-\infty}^{\infty} W_-(-u)e^{-j2\pi ut}\, du$$

$$= \int_{-\infty}^{\infty} [W_+(u)]^* e^{-j2\pi ut}\, du.$$

Therefore,

$$\int_{-\infty}^{\infty} W_-(f)e^{+j2\pi ft}\, df = \left\{\int_{-\infty}^{\infty} W_+(u)e^{+j2\pi ut}\, du\right\}^*. \qquad (4.38)$$

Because $z + z^* = 2\,\mathrm{Re}\{z\}$ for an arbitrary complex number z, (4.37) and (4.38) imply that

$$w(t) = 2\,\mathrm{Re}\left\{\int_{-\infty}^{\infty} W_+(u)e^{+j2\pi ut}\, du\right\}. \qquad (4.39)$$

Now define the frequency function Y by $Y(f) = 2W_+(f + f_c)$. For the frequency function W illustrated in Figures 4–3 and 4–4, the corresponding frequency function Y is as shown in Figure 4–5. It should be clear from the definition of Y that $Y(f) = 0$ for $|f| \geq B$ and $2W_+(u) = Y(u - f_c)$ for $-\infty < u < \infty$.

It follows from (4.39) that

$$w(t) = \mathrm{Re}\left\{\int_{-\infty}^{\infty} Y(u - f_c)e^{+j2\pi ut}\, du\right\}. \qquad (4.40)$$

By substituting f for $u - f_c$, we can rewrite (4.40) as

$$w(t) = \mathrm{Re}\left\{\int_{-\infty}^{\infty} Y(f)\exp\{j2\pi(f + f_c)t\}\, df\right\}.$$

The term $\exp\{j2\pi f_c t\}$ does not depend on f, so

$$w(t) = \mathrm{Re}\left\{\int_{-\infty}^{\infty} Y(f)\exp\{j2\pi ft\}\, df\,\exp\{j2\pi f_c t\}\right\}. \qquad (4.41)$$

But the integral that appears in (4.41) is just the inverse Fourier transform integral for the frequency function Y. Thus, if y is the inverse transform of Y, then

$$w(t) = \mathrm{Re}\{y(t)\exp\{j2\pi f_c t\}\}. \qquad (4.42)$$

From (4.42), we obtain

$$w(t) = [\mathrm{Re}\{y(t)\}]\cos(2\pi f_c t) - [\mathrm{Im}\{y(t)\}]\sin(2\pi f_c t). \qquad (4.43)$$

The expression we seek for the time-domain representation for the frequency function W follows from (4.43).

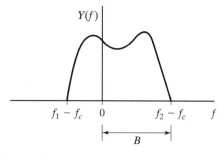

FIGURE 4–5 The frequency function $Y(f) = 2W_+(f + f_c)$.

Recall that our goal is to express $w(t)$ in the form

$$w(t) = v_1(t) \cos(2\pi f_c t) + v_2(t) \sin(2\pi f_c t), \qquad (4.44)$$

where v_1 and v_2 are baseband functions with Fourier transforms that are identically zero for $|f| \geq B$. The functions v_1 and v_2 needed for this representation can be obtained from the real and imaginary parts of $y(t)$. If

$$v_1(t) = \text{Re}\{y(t)\} \qquad (4.45a)$$

and

$$v_2(t) = -\text{Im}\{y(t)\}, \qquad (4.45b)$$

then (4.43) is equivalent to (4.44). It remains to show that v_1 and v_2 are baseband functions with the desired bandwidth.

Before dealing with the bandwidth issue for v_1 and v_2, it is worthwhile to discuss the function y that is used to obtain those functions. It may seem at first glance that y ought to be a real function. That this is not true in general is illustrated by Figure 4–5. Notice that $Y(f)$ does not have conjugate symmetry in this example, so its inverse Fourier transform cannot be a real function. Even though $W(-f) = [W(f)]^*$ in this example, it is not true that $Y(-f) = [Y(f)]^*$. For some band-pass frequency functions, such as the one in the next example, no choice of the frequency f_c gives $Y(-f) = [Y(f)]^*$. For certain other band-pass frequency functions, f_c can be selected to give $Y(-f) = [Y(f)]^*$. (See Problem 4.12.)

Example 4–3 A Band-pass Frequency Function

Consider the ideal band-pass frequency function defined by

$$W(f) = \begin{cases} (|f| - 500)/750, & 500 < |f| < 1250, \\ -(|f| - 1500)/250, & 1250 < |f| < 1500, \\ 0, & \text{otherwise.} \end{cases}$$

This frequency function is illustrated in Figure 4–6, along with the corresponding frequency function Y. Clearly, no choice of f_c leads to a function Y that has conjugate symmetry. ∎

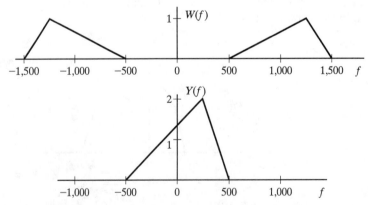

FIGURE 4–6 An example of a band-pass frequency function.

The next step is to show that the Fourier transforms of v_1 and v_2 are identically zero for $|f| \geq B$. Recall from (4.45a) that v_1 is obtained from y by extracting the real part:

$$v_1(t) = \text{Re}\{y(t)\}.$$

Because $z + z^* = 2\,\text{Re}\{z\}$ for an arbitrary complex number z,

$$v_1(t) = \frac{\{y(t) + y^*(t)\}}{2}. \tag{4.46}$$

But

$$y(t) = \int_{-\infty}^{\infty} Y(f)e^{+j2\pi ft}\,df \tag{4.47}$$

and

$$y^*(t) = \left[\int_{-\infty}^{\infty} Y(f)e^{+j2\pi ft}\,df\right]^*$$

$$= \int_{-\infty}^{\infty} Y^*(f)e^{-j2\pi ft}\,df.$$

It follows from the change of variable $u = -f$ that

$$y^*(t) = \int_{-\infty}^{\infty} Y^*(-u)e^{+j2\pi ut}\,du. \tag{4.48}$$

From (4.46)–(4.48), it follows that

$$v_1(t) = \frac{1}{2}\int_{-\infty}^{\infty}\{Y(f) + Y^*(-f)\}e^{+j2\pi ft}\,df. \tag{4.49}$$

Equation (4.49) implies that the Fourier transform of v_1 is given by

$$V_1(f) = \frac{Y(f) + Y^*(-f)}{2}. \tag{4.50a}$$

But $Y(f) = 0$ for $|f| \geq B$, so $Y^*(-f) = 0$ for $|f| \geq B$ as well. It then follows from (4.50a) that $V_1(f) = 0$ for $|f| \geq B$.

A similar development shows that the Fourier transform of v_2 is

$$V_2(f) = \frac{j\{Y(f) - Y^*(-f)\}}{2}, \tag{4.50b}$$

which implies that $V_2(f) = 0$ for $|f| \geq B$. Thus, the functions v_1 and v_2 have Fourier transforms that are identically zero for $|f| \geq B$, as we set out to show.

Nearly all applications of band-pass frequency functions satisfy $B < f_c$. In fact, for most applications, $B \ll f_c$. For example, a typical VHF radio transmission system for digitized voice signals might have a bandwidth ($2B$) of 25 kHz and employ a carrier frequency of 60 MHz, in which case B would be more than three orders of magnitude smaller than f_c.

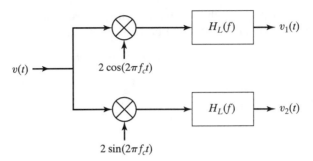

FIGURE 4–7   Generation of the baseband signals v_1 and v_2.

Suppose that v is an ideal band-pass signal with absolute bandwidth not greater than $2B$. We have shown that such a signal can be expressed as

$$v(t) = v_1(t) \cos(2\pi f_c t) + v_2(t) \sin(2\pi f_c t),$$

where v_1 and v_2 are baseband functions whose Fourier transforms are identically zero for $|f| \geq B$. So far we have given only mathematical descriptions of v_1 and v_2. However, it is easy to show that if $B < f_c$, then v_1 and v_2 are baseband signals that can be generated by the system illustrated in Figure 4–7. The system consists of two multipliers and two low-pass filters, and the only need for the filters is to remove double-frequency components at the outputs of the multipliers. (See Problem 4.14 for further details.)

Recall that for a band-pass frequency function W that has conjugate symmetry, the frequency function Y is defined by $Y(f) = 2W_+(f + f_c)$, $-\infty < f < \infty$. There are band-pass frequency functions for which Y has conjugate symmetry for some choice of the frequency f_c. For Y to have such conjugate symmetry, the band-pass frequency function W must have an additional property known as local symmetry. A band-pass frequency function W is said to have *local symmetry about* f_c if, for each real number δ,

$$[W_+(f_c + \delta)]^* = W_+(f_c - \delta), \quad -\infty < f < \infty.$$

This condition is illustrated in Figure 4–8. It is easy to show that it is equivalent to $[W_-(-f_c + \delta)]^* = W_-(-f_c - \delta)$, $-\infty < f < \infty$. Because $Y(f) = 2W_+(f + f_c)$, local symmetry of W about f_c implies that $[Y(f)]^* = Y(-f)$ for all f. Thus, if the band-pass frequency function W has both local symmetry about f_c and conjugate symmetry, the function Y has conjugate symmetry.

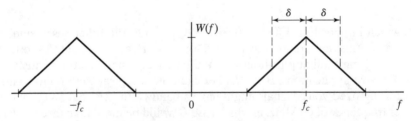

FIGURE 4–8   An ideal band-pass frequency function with local symmetry.

If W has local symmetry about f_c, the time-domain representation of W is simpler than it is for a general band-pass frequency function. The local symmetry of W guarantees that Y has conjugate symmetry, and this in turn guarantees that y, the inverse Fourier transform of Y, is a real function. It follows that $\mathrm{Im}\{y(t)\}$ is identically zero, so (4.44) reduces to

$$w(t) = v_1(t)\cos(2\pi f_c t).$$

Because $y(t)$ is real,

$$v_1(t) = \mathrm{Re}\{y(t)\} = y(t), \quad -\infty < t < \infty.$$

As a result,

$$w(t) = y(t)\cos(2\pi f_c t). \tag{4.51}$$

The conclusion is that if W is a locally symmetric ideal band-pass frequency function, the function y is real and the representation of w is given by (4.51).

If the locally symmetric band-pass frequency function is the *transfer function* $H(f)$ for a linear time-invariant filter, we follow the convention that the impulse response of the filter is written as

$$h(t) = 2\,g(t)\cos(2\pi f_c t). \tag{4.52}$$

The inclusion of the factor 2 in (4.52) leads to several notational advantages in the results that follow. If $h(t)$ and $g(t)$ are related as in that equation and h is the impulse response of a locally symmetric band-pass filter, the filter with impulse response g is referred to as the *baseband equivalent* of the band-pass filter. It follows from the modulation theorem of Fourier transforms that

$$H(f) = G(f - f_c) + G(f + f_c), \quad -\infty < f < \infty, \tag{4.53}$$

and it follows from (4.53) that

$$G(f) = H_+(f + f_c), \quad -\infty < f < \infty.$$

Consider the problem of evaluating the output of an ideal band-pass filter if the input is a carrier-modulated signal. The evaluation is simplified greatly if the signal is an ideal band-pass signal and the transfer function of the filter has local symmetry. Suppose that the input signal is

$$w(t) = v_1(t)\cos(2\pi f_c t) + v_2(t)\sin(2\pi f_c t).$$

It should be clear from the definition of an ideal band-pass signal that, regardless of the transfer function of the filter, if the input to the filter is an ideal band-pass signal, the output must also be an ideal band-pass signal. It follows that the output can be written as

$$\hat{w}(t) = w_1(t)\cos(2\pi f_c t) + w_2(t)\sin(2\pi f_c t), \tag{4.54}$$

for some baseband functions w_1 and w_2. If the filter impulse response is given by (4.52), these baseband functions are just the results of passing the signals v_1 and v_2 through the baseband equivalent filter. That is, $w_1 = v_1 * g$ and $w_2 = v_2 * g$. As a result, the output of a locally symmetric band-pass filter can be determined by working with the baseband functions v_1, v_2, and g. The reader may wish to write out the integral expression for convolving w with h in order to gain an appreciation for the amount of labor that is saved by the use of (4.54) together with $w_1 = v_1 * g$ and $w_2 = v_2 * g$.

To derive (4.54), first observe that it follows from the modulation theorem of Fourier transforms that the band-pass signal

$$w(t) = v_1(t) \cos(2\pi f_c t) + v_2(t) \sin(2\pi f_c t)$$

has Fourier transform

$$W(f) = \frac{V_1(f + f_c) + V_1(f - f_c)}{2} + \frac{j\{V_2(f + f_c) - V_2(f - f_c)\}}{2}. \qquad (4.55)$$

The Fourier transform of the output of the filter is $W(f)H(f)$, where

$$H(f) = G(f + f_c) + G(f - f_c)$$

is the transfer function for the filter. Therefore, the output of the filter has terms of the form

$$\{V_i(f + f_c) \pm V_i(f - f_c)\}H(f) = \\ \{V_i(f + f_c) \pm V_i(f - f_c)\}\{G(f + f_c) + G(f - f_c)\}, \quad (4.56a)$$

or, equivalently,

$$\{V_i(f + f_c) \pm V_i(f - f_c)\}H(f) = \\ V_i(f + f_c)G(f + f_c) \pm V_i(f - f_c)G(f - f_c), \quad (4.56b)$$

which implies

$$\{V_i(f + f_c) \pm V_i(f - f_c)\}H(f) = W_i(f + f_c) \pm W_i(f - f_c), \qquad (4.56c)$$

where $W_i(f) = V_i(f)G(f)$ for $-\infty < f < \infty$ and for $i = 1$ and $i = 2$. To see that (4.56b) is true, notice that if the right-hand side of (4.56a) is expanded, it has four terms. An examination of these terms shows that two of them are identically zero. For example, $V_1(f + f_c)G(f - f_c) = 0$ for all f, as illustrated in Figure 4–9.

More generally, for each value of i,

$$V_i(f + f_c)G(f - f_c) = 0, \quad -\infty < f < \infty \qquad (4.57a)$$

and

$$V_i(f - f_c)G(f + f_c) = 0, \quad -\infty < f < \infty. \qquad (4.57b)$$

It follows that

$$W(f)H(f) = \frac{W_1(f + f_c) + W_1(f - f_c)}{2} + \frac{j\{W_2(f + f_c) - W_2(f - f_c)\}}{2}.$$

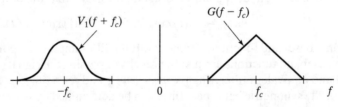

FIGURE 4–9   Illustration of the fact that $V_1(f + f_c)G(f - f_c) = 0$.

Thus, the output signal $\hat{w}(t)$ has Fourier transform

$$\hat{W}(f) = W(f)H(f)$$
$$= \frac{W_1(f + f_c) + W_1(f - f_c)}{2} + \frac{j\{W_2(f + f_c) - W_2(f - f_c)\}}{2}, \quad (4.58)$$

where $W_i(f) = V_i(f)G(f)$ for $-\infty < f < \infty$.

Now let w_i be the inverse Fourier transform of W_i. From the modulation theorem of Fourier transforms, we know that, because of (4.58), $\hat{W}(f)$ corresponds to the time-domain signal

$$\hat{w}(t) = w_1(t)\cos(2\pi f_c t) + w_2(t)\sin(2\pi f_c t).$$

The frequency-domain representation

$$W_i(f) = V_i(f)G(f)$$

corresponds to $w_i = v_i * g$ in the time domain. Hence, we have shown that

$$\hat{w}(t) = w_1(t)\cos(2\pi f_c t) + w_2(t)\sin(2\pi f_c t),$$

where $w_i = v_i * g$ for $i = 1$ and $i = 2$.

Even if the ideal band-pass filter does not have local symmetry about f_c, (4.54) is still valid, and it can be shown that four convolutions involving baseband functions give the necessary results for determining w_1 and w_2. The derivation of the result for locally symmetric band-pass filters can be generalized easily to provide the more general result for ideal band-pass filters that do not have local symmetry.

4.6 BAND-PASS RANDOM PROCESSES

One of the applications of the time-domain representations derived in the previous section is to the representation of the autocorrelation function for a random process whose spectral density is a band-pass frequency function. Let S_X be such a spectral density function for a zero-mean, wide-sense stationary random process $X(t)$, and apply the results of Section 4.5 by letting

$$W(f) = S_X(f), \quad -\infty < f < \infty.$$

Thus, we can write $S_X(f) = W_+(f) + W_-(f)$, where $W_+(f)$ is the positive part of the spectral density and $W_-(f)$ is the negative part. Assume, as illustrated in Figure 4–10, that the absolute bandwidth of the spectral density function is not greater than $2B$. Throughout this section, it is also assumed that $B \leq f_c$.

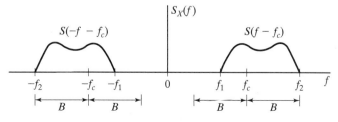

FIGURE 4–10   A band-pass spectral density function.

The situation is somewhat simpler than the general application of the results of Section 4.5, because the spectral density function must be real and even; that is, $S_X(f)$ is real for each f, and $S_X(f) = S_X(-f), -\infty < f < \infty$. It follows that the positive and negative parts of W are real functions that are related by $W_+(f) = W_-(-f)$ for $-\infty < f < \infty$. Next, define $S(f) = W_+(f + f_c)$ for $-\infty < f < \infty$, and observe that this implies that $S(f - f_c) = W_+(f)$. But since $W_+(f) = W_-(-f)$, we have $W_-(f) = S(-f - f_c)$. As a result, the band-pass spectral density can be written as

$$S_X(f) = S(f - f_c) + S(-f - f_c), \tag{4.59}$$

where the frequency function S satisfies $S(f) = 0$ for $|f| \geq B$. This situation is illustrated in Figure 4–10. It is not true in general that $S(f) = S(-f)$. Of course, if the spectral density S_X has *local symmetry* about f_c, then S is a symmetrical function.

The spectral density function S_X is an ideal band-pass frequency function, so the results of Section 4.5 imply that the inverse Fourier transform of S_X—the autocorrelation function for $X(t)$—can be expressed as

$$R_X(\tau) = f_1(\tau) \cos(2\pi f_c \tau) + f_2(\tau) \sin(2\pi f_c \tau), \tag{4.60}$$

where f_1 and f_2 have Fourier transforms that are zero for $|f| \geq B$.

The function Y defined in Section 4.5 is given by

$$Y(f) = 2W_+(f + f_c) = 2S(f),$$

so, from (4.41),

$$R_X(\tau) = 2 \, \text{Re}\left\{ \int_{-\infty}^{\infty} S(f) \exp\{j2\pi f \tau\} \, df \, \exp\{j2\pi f_c \tau\} \right\}. \tag{4.61}$$

It follows that

$$f_1(\tau) = 2 \int_{-\infty}^{\infty} S(f) \cos(2\pi f \tau) \, df \tag{4.62a}$$

and

$$f_2(\tau) = -2 \int_{-\infty}^{\infty} S(f) \sin(2\pi f \tau) \, df. \tag{4.62b}$$

If S_X has local symmetry about f_c, then $S(f) = S(-f)$ for all f, and it is clear from (4.62b) that $f_2(\tau)$ is identically zero. In this special case, (4.60) simplifies to

$$R_X(\tau) = f_1(\tau) \cos(2\pi f_c \tau).$$

Of much greater interest for applications in communications systems analysis is the fact that the random process itself can be expressed in the form of the signal representation of (4.44). It can be shown that if the zero-mean random process $X(t)$ is wide-sense stationary and has an ideal band-pass spectral density, it can be expressed as

$$X(t) = X_1(t) \cos(2\pi f_c t) + X_2(t) \sin(2\pi f_c t), \tag{4.63}$$

where $X_1(t)$ and $X_2(t)$ are zero-mean, jointly wide-sense stationary random processes with spectral densities that are identically zero for $|f| \geq B$. Some consequences of this representation in terms of jointly wide-sense stationary random processes are given in the next exercise.

Exercise 4–4. Show that (4.63) implies that

$$R_1(\tau) = R_2(\tau), \quad -\infty < \tau < \infty \tag{4.64a}$$

and

$$R_{21}(\tau) = -R_{12}(\tau), \quad -\infty < \tau < \infty. \tag{4.64b}$$

Solution. The autocorrelation function for $X(t)$ is given by

$$R_X(\tau) = E\{X(t + \tau)X(t)\}.$$

Substituting from (4.63), expanding the product, taking the expectation term by term, and employing standard trigonometric identities, we obtain

$$
\begin{aligned}
R_X(\tau) = {}& \left\{ \frac{R_1(\tau) + R_2(\tau)}{2} \right\} \cos(2\pi f_c \tau) \\
& + \left\{ \frac{R_{21}(\tau) - R_{12}(\tau)}{2} \right\} \sin(2\pi f_c \tau) \\
& + \left\{ \frac{R_1(\tau) - R_2(\tau)}{2} \right\} \cos(4\pi f_c t + 2\pi f_c \tau) \\
& + \left\{ \frac{R_{21}(\tau) + R_{12}(\tau)}{2} \right\} \sin(4\pi f_c t + 2\pi f_c \tau),
\end{aligned} \tag{4.65}
$$

where

$$R_i(\tau) = E\{X_i(t + \tau)X_i(t)\}$$

and

$$R_{ij}(\tau) = E\{X_i(t + \tau)X_j(t)\}$$

for $i = 1, 2$ and $j = 1, 2$. Because $X(t)$ is wide-sense stationary, the last two terms of (4.65) must be constant. (There can be no dependence on t.) The only way for the third term to be constant is for $R_1(\tau) - R_2(\tau)$ to be zero for all τ, and the only way for the fourth term to be constant is for $R_{21}(\tau) + R_{12}(\tau)$ to be zero for each τ. But these observations imply that (4.64) must hold. ∎

Notice that it follows from (4.64) and (4.65) that

$$R_X(\tau) = R_1(\tau) \cos(2\pi f_c \tau) + R_{21}(\tau) \sin(2\pi f_c \tau), \tag{4.66}$$

from which the functions f_1 and f_2 of (4.60) can be identified, namely, $f_1 = R_1$ and $f_2 = R_{21}$. Thus, we can write (4.62a) as

$$R_1(\tau) = 2 \int_{-\infty}^{\infty} S(f) \cos(2\pi f \tau) \, df \tag{4.67a}$$

and (4.62b) as

$$R_{21}(\tau) = -2 \int_{-\infty}^{\infty} S(f) \sin(2\pi f \tau) \, df. \tag{4.67b}$$

One important observation that follows from (4.66) is that the function f_1 in (4.60) is an autocorrelation function ($f_1 = R_1$), but the function f_2 is not. We see that $f_2 = R_{21}$, which is a crosscorrelation function. Hence, there is no guarantee that $f_2(-\tau) = f_2(\tau)$, because crosscorrelation functions need not have such symmetry. In fact, it turns out that $f_2(-\tau) = f_2(\tau)$ only in the trivial situation in which $f_2(\tau) = 0$ for all τ.

To see this, observe from the definitions of the two crosscorrelation functions that

$$R_{21}(-\tau) = R_{12}(\tau).$$

Together with (4.64b), this equation implies that

$$R_{21}(-\tau) = -R_{21}(\tau). \tag{4.68}$$

Equation (4.68) can also be derived from (4.67b) by using the fact that

$$\sin(-2\pi f \tau) = -\sin(2\pi f \tau).$$

As a result of (4.68), the only way for R_{21} to satisfy $R_{21}(-\tau_0) = R_{21}(\tau_0)$ for a particular τ_0 is if $-R_{21}(\tau_0) = R_{21}(\tau_0)$, and this is true only if $R_{21}(\tau_0) = 0$. Therefore, $R_{21}(-\tau_0) = R_{21}(\tau_0)$ only if the random processes $X_1(t)$ and $X_2(t)$ are such that

$$E\{X_1(t + \tau_0)X_2(t)\} = 0$$

for all t. Furthermore,

$$R_{21}(-\tau) = R_{21}(\tau), \quad -\infty < \tau < \infty$$

only if

$$E\{X_1(t + \tau)X_2(t)\} = 0$$

for all t and all τ (i.e., the two zero-mean random processes are uncorrelated).

For the special case in which $X_1(t)$ and $X_2(t)$ are uncorrelated random processes, $R_{21}(\tau) = 0$ for all τ, and the autocorrelation for $X(t)$ can be written as

$$R_X(\tau) = R_1(\tau) \cos(2\pi f_c \tau). \tag{4.69}$$

From the modulation theorem of Fourier transforms, it follows that the spectral density for $X(t)$ is given by

$$S_X(f) = S(f - f_c) + S(f + f_c), \tag{4.70}$$

where S is one-half the Fourier transform of R_1. As the Fourier transform of a real function, S must be an even function, so S_X has local symmetry about f_c. Also, since S is an even function, $S(f + f_c) = S(-f - f_c)$. It follows that if $X_1(t)$ and $X_2(t)$ are uncorrelated, (4.70) agrees with (4.59), and the function S in the former equation is one-half the Fourier transform of R_1. Note, however, that (4.59) is valid for any bandpass random process, but (4.70) is valid only if $X_1(t)$ and $X_2(t)$ are uncorrelated random processes.

Although it is not true in general that $R_{21}(\tau) = 0$ for all τ, it is true in general that $R_{21}(0) = 0$. This follows from (4.67b) by setting $\tau = 0$. The implication is that, for any choice of t_0, $E\{X_2(t_0)X_1(t_0)\} = 0$. Thus, although the two random *processes* are not uncorrelated, the two random *variables* $X_1(t_0)$ and $X_2(t_0)$ *are* uncorrelated for any choice of the sampling time t_0.

Because R_1 is the autocorrelation function for the random process $X_1(t)$, the Fourier transform of R_1, denoted by S_1, is the spectral density function for $X_1(t)$. Because

$R_2 = R_1$, S_1 is also the spectral density function for the random process $X_2(t)$. Let S_{21} be the Fourier transform of the crosscorrelation function R_{21}; that is,

$$S_{21}(f) = \int_{-\infty}^{\infty} R_{21}(\tau)e^{-j2\pi f\tau} d\tau. \tag{4.71}$$

The function S_{21} is known as the *cross-spectral density function* for the two random processes $X_1(t)$ and $X_2(t)$. As shown in the next exercise, the spectral density for $X(t)$ can be expressed in terms of the functions S_1 and S_{21}.

Exercise 4–5. Let $X(t)$ be a zero-mean, wide-sense stationary random process with ideal band-pass spectral density. Use the representation given by (4.63) and the modulation theorem of Fourier transforms to find the spectral density for $X(t)$ in terms of the spectral density for the random process $X_1(t)$ and the cross-spectral density for the random processes $X_1(t)$ and $X_2(t)$.

Solution. The general form of the modulation theorem of Fourier transforms, which is developed in Problem 4.6, gives the Fourier transforms of $v(t) \cos(2\pi f_c t)$ and $v(t) \sin(2\pi f_c t)$ in terms of the Fourier transform of $v(t)$. Applying these results shows that the Fourier transform of $R_1(\tau) \cos(2\pi f_c \tau)$ is

$$\frac{S_1(f - f_c) + S_1(f + f_c)}{2}$$

and the Fourier transform of $R_{21}(\tau) \sin(2\pi f_c \tau)$ is

$$-\frac{j[S_{21}(f - f_c) - S_{21}(f + f_c)]}{2}.$$

Combining the two terms, we find that the spectral density for $X(t)$ is given by

$$S(f) = \frac{S_1(f - f_c) + S_1(f + f_c)}{2} - \frac{j[S_{21}(f - f_c) - S_{21}(f + f_c)]}{2}. \qquad \blacksquare$$

We know from previous sections in this chapter that S_1 and S_{21} have conjugate symmetry, because they are Fourier transforms of the real functions R_1 and R_{21}, respectively. In addition, because $R_1(-\tau) = R_1(\tau)$ for all τ, S_1 is an even function. However, according to (4.68),

$$R_{21}(-\tau) = -R_{21}(\tau)$$

for all τ. The complex conjugate of the cross-spectral density is given by

$$[S_{21}(f)]^* = \int_{-\infty}^{\infty} R_{21}(\tau)e^{+j2\pi f\tau} d\tau = \int_{-\infty}^{\infty} R_{21}(-u)e^{-j2\pi fu} du$$

$$= \int_{-\infty}^{\infty} -R_{21}(u)e^{-j2\pi fu} du = -S_{21}(f). \tag{4.72}$$

Just as $z^* = z$ implies that the complex number z is real, $z^* = -z$ implies that z is purely imaginary. As demonstrated in (4.72), the cross-spectral density for the random processes $X_1(t)$ and $X_2(t)$ is purely imaginary. As mentioned in the first sentence of this paragraph, S_{21} has conjugate symmetry; that is,

$$[S_{21}(f)]^* = S_{21}(-f), \quad -\infty < f < \infty.$$

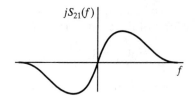

FIGURE 4–11   Cross-spectral density function S_{21}.

This fact, together with (4.72), implies that

$$S_{21}(-f) = -S_{21}(f)$$

for each f. It follows that $jS_{21}(f)$ is real and is an odd function of f; that is,

$$jS_{21}(-f) = -jS_{21}(f), \quad -\infty < f < \infty,$$

as illustrated in Figure 4–11.

The random processes $X_1(t)$ and $X_2(t)$ can be generated from $X(t)$ by the use of the system shown in Figure 4–12. The system is identical to the one employed for deterministic signals in Section 4.5.2. (See Figure 4–7.) This approach is not used in the analytical development of the properties of the correlation functions and spectral densities, largely because the system of Figure 4–12 is time varying. In particular, the inputs to the low-pass filters are not wide-sense stationary random processes. As a result, spectral analysis methods cannot be applied to the filtering operations performed in that system.

The derivation of the representation of a band-pass random process is typically given in terms of the Hilbert transform (e.g., see [4.1] and [4.2]). An alternative approach is given in [4.3]. Even though the system of Figure 4–12 may be difficult to analyze, it generates random processes $X_1(t)$ and $X_2(t)$ that are equivalent to those obtained by the use of the Hilbert transform. (See [4.1] or [4.4].) The outputs $X_1(t)$ and $X_2(t)$ are wide-sense stationary, even if the intermediate processes generated within the system are not; furthermore, $X_1(t)$ and $X_2(t)$ have the correlation functions and spectral densities described in this section.

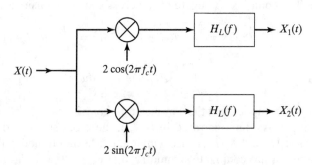

FIGURE 4–12   Decomposition of the ideal band-pass random process $X(t)$.

One property that is obvious from Figure 4–12 is very important: If $X(t)$ is a Gaussian random process, then $X_1(t)$ and $X_2(t)$ are jointly Gaussian. Even though the system shown is not time invariant, it is linear, and that is all that is required to guarantee that the output processes are jointly Gaussian. If the spectral density S_X for the Gaussian random process $X(t)$ is such that $R_{21}(\tau) = 0$ for all τ, then the output processes are uncorrelated. Because they are jointly Gaussian, the output processes are also independent. In general, however, $R_{21}(\tau)$ is not zero for all τ, and the random processes $X_1(t)$ and $X_2(t)$ are statistically dependent.

REFERENCES AND SUGGESTIONS FOR FURTHER READING

[4.1] A. Papoulis, *Probability, Random Variables, and Stochastic Processes*, 3rd ed., New York: McGraw-Hill, 1991.

[4.2] B. P. Lathi, *An Introduction to Random Signals and Communication Theory*, Scranton, PA: International, 1968.

[4.3] A. J. Viterbi, *Principles of Coherent Communication*, New York: McGraw-Hill, 1966.

[4.4] H. Stark and J. W. Woods, *Probability, Random Processes, and Estimation Theory for Engineers*, 2nd ed., Englewood Cliffs, NJ: Prentice Hall, 1994.

PROBLEMS

4.1 Derive the spectral density in item (6) of Table 4–1 in two ways. First use items (4) and (5) from the table and the fact that multiplication in the time domain corresponds to convolution in the frequency domain. (Watch out for the constants.) Second, use item (4) and the modulation theorem of Fourier transforms, which states that if $v(t)$ has Fourier transform $V(\omega)$, then the Fourier transform of the signal

$$w(t) = v(t)\cos(\omega_c t)$$

is

$$W(\omega) = \frac{V(\omega - \omega_c) + V(\omega + \omega_c)}{2}.$$

The proof of the modulation theorem is outlined in Problem 4.6.

4.2 Consider the signal $v(t) = 1/\sqrt{T}$ for $-T/2 < t < +T/2$ and $v(t) = 0$ otherwise. The Fourier transform of this signal is

$$V(f) = \sqrt{T}\,\text{sinc}(fT).$$

Show how this fact can be used to derive item (1) in Table 4–1. [*Hint:* The convolution of $v(t)$ with itself is a triangle, and convolution in the time domain corresponds to multiplication in the frequency domain.]

4.3 A linear time-invariant filter has transfer function $H(f) = 1$ for $|f| < B$ and $H(f) = 0$ otherwise. The input random process $X(t)$ is a zero-mean, wide-sense stationary, Gaussian random process with autocorrelation function $R_X(\tau) = \beta \exp(-\alpha|\tau|)$. The corresponding output is $Y(t)$.
(a) Find $E\{[Y(t)]^2\}$.
(b) Find $P[|Y(t_0)| \le \gamma]$ for an arbitrary positive value of γ.

4.4 The wide-sense stationary random process $X(t)$ has spectral density $S_X(f) = 1$ for $|f| < B$ and $S_X(f) = 0$ otherwise. Let $Y(t)$ be the output when $X(t)$ is the input to a linear time-invariant filter with transfer function $H(f) = \cos(\pi f/2W)$ for $|f| < W$ and $H(f) = 0$ otherwise. Find the output spectral density $S_Y(f)$, and give the value of $E\{[Y(t)]^2\}$. Consider the two cases $B \geq W$ and $B < W$.

4.5 A time-invariant linear filter has transfer function

$$H(\omega) = \exp(-\omega^2), \quad -\infty < \omega < \infty.$$

The input to this filter is a white-noise process $X(t)$ with power spectral density

$$S_X(\omega) = \frac{N_0}{2}, \quad -\infty < \omega < \infty,$$

and the corresponding output is the random process $Y(t)$. Find the expected value of the instantaneous power in the output process. [*Hint*: To help evaluate the resulting integral, notice that the transfer function has a Gaussian shape. Can you relate it to a probability density function? How might this help solve the problem?]

4.6 (a) Prove the modulation theorem of Fourier transforms by using the following steps: First, note that the signal $v(t)$ has Fourier transform given by

$$V(\omega) = \int_{-\infty}^{\infty} v(t)e^{-j\omega t}\, dt.$$

Next, use the fact that

$$\cos(\omega_c t) = \tfrac{1}{2}\{\exp(+j\omega_c t) + \exp(-j\omega_c t)\}$$

to infer that the Fourier transform of $w_1(t) = v(t)\cos(\omega_c t)$ can be written as

$$W_1(\omega) = \tfrac{1}{2}\int_{-\infty}^{\infty} v(t)\exp[-j(\omega - \omega_c)t]\, dt + \tfrac{1}{2}\int_{-\infty}^{\infty} v(t)\exp[-j(\omega + \omega_c)t]\, dt.$$

By comparing these two integrals with the Fourier transform integral, show that the expression for W_1, is equivalent to

$$W_1(\omega) = \tfrac{1}{2}[V(\omega - \omega_c) + V(\omega + \omega_c)].$$

(b) Repeat the steps in part **(a)** for $w_2(t) = v(t)\cos(\omega_c t + \theta)$, using the identity

$$\cos(\omega_c t + \theta) = \tfrac{1}{2}\{\exp[+j(\omega_c t + \theta)] + \exp[-j(\omega_c t + \theta)]\}.$$

Show that

$$W_2(\omega) = \tfrac{1}{2}[V(\omega - \omega_c)e^{+j\theta} + V(\omega + \omega_c)e^{-j\theta}].$$

Notice that this last result reduces to the result in part **(a)** if $\theta = 0$.

(c) Show that the result in **(b)** applied for the special case $\theta = -\pi/2$ proves that the Fourier transform of $v(t)\sin(\omega_c t)$ is $j[V(\omega + \omega_c) - V(\omega - \omega_c)]/2$.

4.7 Suppose a random process is defined by

$$Z(t) = A_1(t)\cos(\omega_c t + \Theta) + A_2(t)\sin(\omega_c t + \Theta).$$

Assume that Θ is uniformly distributed on $[0, 2\pi]$ and independent of $A_1(t)$ and $A_2(t)$. Suppose further that $A_1(t)$ and $A_2(t)$ are each zero-mean, wide-sense stationary random processes with autocorrelation function $R_A(\tau)$. Suppose also that $A_1(t)$ and $A_2(t)$ are uncorrelated. Find the spectral density for $Z(t)$ in terms of the spectral density $S_A(\omega)$ for $A_1(t)$ and $A_2(t)$. Compare this result with (4.19).

4.8 Let $\zeta(t)$ represent the waveform for a data pulse, as in Section 4.4.

(a) Suppose that a communication signal is given by

$$x(t) = \sqrt{2}\,\zeta(t)\cos(\omega_c t).$$

What is the energy spectrum $|X(\omega)|^2$? Give sufficient conditions for the shape of this energy spectrum to be identical to the shape of the spectral density $S_Y(\omega)$ given in (4.26). That is, give sufficient conditions for

$$|X(\omega)|^2 = \text{constant} \times S_Y(\omega).$$

(b) Consider the signal defined by

$$y(t) = \sqrt{2}\,\zeta(t)\cos(\omega_c t + \theta).$$

The parameter θ is a deterministic constant. What is the energy spectrum for $y(t)$? [*Hint*: Judicious use of the result of Problem 4.6(b) may be helpful.]

4.9 Consider the random signal given by $A(t) = D(t - U),\ -\infty < t < \infty$, where U is a random variable that is uniformly distributed on the interval $[0, T]$ and the random signal $D(t)$ is given by (4.20). Suppose that the random sequence (A_n) is wide-sense stationary, is independent of U, and has autocorrelation function $\rho_k = E\{A_0 A_k\},\ -\infty < k < \infty$.

(a) Show that $A(t)$ is wide-sense stationary and has autocorrelation function given by

$$R_A(\tau) = \sum_{k=-\infty}^{\infty} \rho_k r_\zeta(\tau - kT),$$

where

$$r_\zeta(\tau) = T^{-1} \int_{-\infty}^{\infty} \zeta(t)\zeta(t + \tau)\,dt.$$

(b) Apply the result obtained in part **(a)** to find the autocorrelation and spectral density functions for the random signal $A(t)$ if $\zeta(t) = p_T(t), \rho_0 = \alpha^2, \rho_1 = \rho_{-1} = \beta$, and $\rho_k = 0$ for $|k| \geq 2$.

(c) For $\alpha = 1$ and $\beta = \frac{1}{2}$, compare a plot of the spectral density obtained in part **(b)** with the spectral density that is obtained if the sequence (A_n) is a sequence of independent zero-mean random variables with $E\{A_n^2\} = 1$.

4.10 Which of the following frequency functions are ideal band-pass frequency functions as defined in Section 4.5?

(a) $W(f) = \begin{cases} \sin(\pi f), & 10 < |f| < 15, \\ 0, & \text{otherwise.} \end{cases}$

(b) $W(f) = \exp\{-(f - 100)^2\},\ -\infty < f < \infty.$

(c) $W(f) = \begin{cases} \exp\{j\pi f/50\}, & 1{,}000 < |f| < 2{,}000, \\ 0, & \text{otherwise.} \end{cases}$

Find the cutoff bandwidth for each of the ideal band-pass frequency functions.

4.11 Suppose the functions in Problem 4.10 are transfer functions for time-invariant linear filters. Which of them have well-defined half-power bandwidths? For each that does have a well-defined half-power bandwidth, determine the maximum gain W_m and the half-power bandwidth.

4.12 Consider the ideal band-pass frequency function defined by

$$W(f) = \begin{cases} (|f| - 500)/500, & 500 < |f| < 1000, \\ -(|f| - 1500)/500, & 1000 < |f| < 1500, \\ 0, & \text{otherwise.} \end{cases}$$

This frequency function is illustrated in the following diagram:

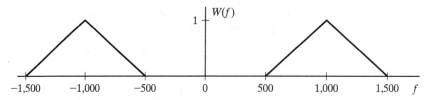

Consider the function y, defined as the inverse transform of

$$Y(f) = 2W_+(f + f_c), -\infty < f < \infty.$$

(a) What choice of f_c makes y a real function?

(b) Suppose $f_c = 750$. Specify $Y(f)$ for $-\infty < f < \infty$ and $y(t)$ for $-\infty < t < \infty$.

4.13 (a) Following the development of the results for V_1 in (4.46) through (4.50a), begin with the fact that $v_2(t) = -\text{Im}\{y(t)\}$, and show that

$$V_2(f) = \frac{j\{Y(f) - Y^*(-f)\}}{2}.$$

(b) Use your result from part **(a)** together with (4.50a) to show that, for both $i = 1$ and $i = 2$,

$$V_i(-f) = V_i^*(f), \quad -\infty < f < \infty.$$

4.14 Let v be an ideal band-pass signal with absolute bandwidth $2B$. Then v can be expressed as in (4.44), where v_1 and v_2 are baseband functions with Fourier transforms that are identically zero for $|f| \geq B$. Assume that $B < f_c$. Consider the system shown in the following diagram:

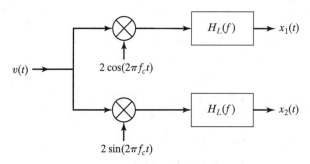

The linear filters in this system are defined by

$$H_L(f) = \begin{cases} 1, & -B \leq f \leq +B \\ 0, & \text{otherwise} \end{cases}.$$

Show that the system generates the signals v_1 and v_2 by demonstrating that $x_1(t) = v_1(t)$ and $x_2(t) = v_2(t)$ for $-\infty < t < \infty$. [*Hint*: It suffices to show that the Fourier transform of x_i is V_i for each i, and (4.50) may be of help in doing so.]

4.15 Let $X(t)$ be an ideal band-pass process that is wide-sense stationary and has autocorrelation function given by

$$R_X(\tau) = f_1(\tau) \cos(2\pi f_c \tau) + f_2(\tau) \sin(2\pi f_c \tau), \quad -\infty < \tau < \infty.$$

Show that if f_1 is an autocorrelation function, f_2 cannot be an autocorrelation function. [*Hint*: Notice that R_X, f_1, and the cosine are even functions, but the sine is an odd function.]

4.16 A wide-sense stationary, zero-mean, Gaussian random process $X(t)$ is the input to the following system:

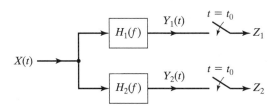

The spectral density for $X(t)$ is given by $S(f) = \alpha \exp(-\beta f^2)$ for $-\infty < f < \infty$. The outputs $Y_1(t)$ and $Y_2(t)$ of the filters are sampled at time t_0 to give the random variables Z_1 and Z_2. The linear time-invariant filters $H_1(f)$ and $H_2(f)$ are ideal band-pass filters with center frequencies f_1 and f_2, respectively, and each has bandwidth $2W$. Assume that $H_i(f) = 1$ for frequencies in the passband and that $0 < W < f_1 < f_2 - 2W$, so that the passbands of the filters do not overlap. In parts **(a)–(e)** and **(g)**, evaluate all integrals, and give your answers in terms of the function Φ and the parameters $\alpha, \beta, t_0, W, f_1$, and f_2 only.

(a) What constraints must be placed on α and β to guarantee that $S(f)$ is a valid spectral density for a wide-sense stationary random process with power P if $0 < P < \infty$? Assume that α and β satisfy these constraints in parts **(b)–(h)**.

(b) Give an expression for P, the power in the random process $X(t)$.

(c) Give an expression for $S_1(f)$, the spectral density for $Y_1(t)$.

(d) Find $m_1 = E\{[Y_1(t_0)]^2\}$ and $m_2 = E\{[Y_2(t_0)]^2\}$.

(e) Give expressions for $\mu_1 = E\{Z_1\}$ and $\mu_2 = E\{Z_2\}$.

(f) Give an integral expression (with a single or double integral) for the crosscorrelation function $R_{1,2}(\tau) = E\{Y_1(t + \tau)Y_2(t)\}$ in terms of the impulse responses of the two filters and the autocorrelation function for the random process $X(t)$ (i.e., your answer should be an integral expression involving the functions h_1, h_2, and R_X). Give an equivalent expression in terms of one or two convolutions.

(g) Give an expression for $E\{Z_1 Z_2\}$. [*Hint*: It may be best to first relate $E\{Z_1 Z_2\}$ to the crosscorrelation function $R_{1,2}$ defined in part **(f)** and then use frequency-domain methods to evaluate the result.]

(h) Find $P(Z_2 > Z_1 - 1)$. Express your answer in terms of Φ and the parameters m_1, m_2, μ_1, and μ_2 only. Explain how the solution to part **(g)** is used to obtain your answer.

Index